IOP Series in Global Health and Radiation Oncology

Series editor
Wilfred Ngwa
Dana-Farber/Harvard Cancer Center, University of Massachusetts Lowell, Johns Hopkins University, USA

Editor biography
Wilfred Ngwa is the Director of the Global Health Catalyst, a cross-institutional collaboration initiative launched at Harvard to catalyse high impact collaborations in global health. He currently serves as Adjunct Professor at the University of Massachusetts, as Associate Professor of Radiation Oncology at Johns Hopkins University and as ICTU Distinguished Professor of Public Health. He is a chair of the Lancet Oncology Commission for Sub-Saharan Africa, has published three books on global health and serves on the editorial board of a number of journals, including ASCO's *Journal of Global Oncology, Ecancermedicalsciences*, and *Frontiers in Oncology*. He has won many awards from Harvard, the USA National Institutes of Health, and international professional organisations for his innovations and leading work in global health to address disparities in the USA and globally.

Aims and scope
This series includes books in the emerging area of global radiation oncology and its applications in global health. Building on the published book by the series editor entitled *Emerging Models for Global Health in Radiation Oncology*, it will further detail the work being done globally to promote cancer research and awareness, particularly in lower-income countries.

A full list of titles published in this series can be found here: https://iopscience.iop.org/bookListInfo/iop-series-in-cancer-research-for-global-radiation-oncology#series.

3D Printing in Radiation Therapy

Edited by
Tanya Kairn
Cancer Care Services, Royal Brisbane and Women's Hospital, Metro North Hospital and Health Service, Herston, Queensland, Australia

Scott B Crowe
Cancer Care Services, Royal Brisbane and Women's Hospital, Metro North Hospital and Health Service, Herston, Queensland, Australia

Tomas Kron
Physical Sciences Department, Peter MacCallum Cancer Centre, Melbourne, Victoria, Australia

IOP Publishing, Bristol, UK

ISBN 978-0-7503-3907-0 (ebook)
ISBN 978-0-7503-3905-6 (print)
ISBN 978-0-7503-3908-7 (myPrint)
ISBN 978-0-7503-3906-3 (mobi)

DOI 10.1088/978-0-7503-3907-0

Version: 20231201

IOP ebooks

British Library Cataloguing-in-Publication Data: A catalogue record for this book is available from the British Library.

Published by IOP Publishing, wholly owned by The Institute of Physics, London

IOP Publishing, No.2 The Distillery, Glassfields, Avon Street, Bristol, BS2 0GR, UK

US Office: IOP Publishing, Inc., 190 North Independence Mall West, Suite 601, Philadelphia, PA 19106, USA

Cover image: Example of electron beam compensator effect from 3D printed radiation bolus. Image credit: Tanya Kairn, Samuel C Peet.

Contents

Editor biographies

Tanya Kairn

Professor Tanya Kairn (PhD) is the director of medical physics for cancer care services at the Royal Brisbane and Women's Hospital, Metro North Hospital and Health Service, Brisbane, Australia. She has published over 150 papers in refereed journals, including 20 papers on applications of 3D printing in radiation therapy. The use of novel 3D-printing techniques for the benefit of cancer patients remains an important focus of Professor Kairn's research.

Scott B Crowe

Professor Scott B Crowe (PhD) is the senior research medical physicist at the Royal Brisbane and Women's Hospital, Metro North Hospital and Health Service, Brisbane, Australia. He is the clinical lead for the Herston Biofabrication Institute Cancer Care Services research programme, and oversaw the process by which the Royal Brisbane and Women's Hospital (Metro North Hospital and Health Service) became the first hospital in Australia to be listed on the Australian Register of Therapeutic Goods as a point-of-care manufacturer of 3D-printed radiation therapy bolus.

Tomas Kron

Professor Tomas Kron (PhD) is the managing director of the Department of Medical Physics at the Peter MacCallum Cancer Centre, Melbourne, Australia, with responsibility for overseeing a service that has produced approximately 2000 patient-matched 3D-printed radiation therapy devices. Professor Kron has over 600 publications, many of which demonstrate his ongoing interest in developing bespoke phantoms and dosimetry systems, including eight papers on 3D-printing topics in the last three years and an early paper on 3D-printed phantom materials in 2015. Professor Tomas Kron was the originator of this book project.

List of contributors

Dr Amirhossein Asfia
Olympus APAC, Notting Hill, Victoria, Australia

Mr Deepak Basaula
Department of Physical Sciences, Peter MacCallum Cancer Centre, Melbourne, Victoria, Australia

Professor Scott B Crowe
Cancer Care Services, Royal Brisbane and Women's Hospital, Metro North Hospital and Health Service, Herston, Queensland, Australia

Ms Mathilde R Desselle
Herston Biofabrication Insitute, Metro North Hospital and Health Service, Herston, Queensland, Australia

Professor Tanya Kairn
Cancer Care Services, Royal Brisbane and Women's Hospital, Metro North Hospital and Health Service, Herston, Queensland, Australia

Dr Georgio Andrew Katsifis
Liverpool-Macarthur Cancer Centre, Liverpool, UK

Professor Tomas Kron
Department of Physical Sciences, Peter MacCallum Cancer Centre, Melbourne, Victoria, Australia

Professor Martin Leary
School of Engineering, RMIT University, Melbourne, Victoria, Australia

Dr James I Novak
Herston Biofabrication Insitute, Metro North Hospital and Health Service, Herston, Queensland, Australia

Dr Samuel C Peet
Cancer Care Services, Royal Brisbane and Women's Hospital, Metro North Hospital and Health Service, Herston, Queensland, Australia

Ms Emily Simpson-Page
Radiation Oncology Princess Alexandra Hospital Raymond Terrace (ROPART), South Brisbane, Queensland, Australia

Associate Professor Natalka Suchowerska
Institute of Medical Physics, University of Sydney, Sydney, New South Wales, Australia

Dr Rance Tino
University of Texas MD Anderson Cancer Center, Houston, TX, USA

Ms Rachael Wilks
Cancer Care Services, Royal Brisbane and Women's Hospital, Metro North Hospital and Health Service, Herston, Queensland, Australia

Dr Adam Unjin Yeo
Department of Physical Sciences, Peter MacCallum Cancer Centre, Melbourne, Victoria, Australia

Contributor biographies

Amirhossein Asfia (PhD) is a biomedical engineer whose recently completed doctoral research included the development of 3D-printed immobilisation masks for radiation therapy.

Deepak Basaula (MSc) is a senior medical physicist at the Peter MacCallum Cancer Centre in Melbourne, Australia, with expertise in implementation and quality assurance of 3D printing, as well as surface guided radiation therapy (SGRT). He also has a strong passion for clinical research, patient safety, and quality improvements in radiation therapy, and the standardisation of radiation oncology medical physics activity.

Mathilde R Desselle (MEng) is the general manager and principal biomedical engineer of the Herston Biofabrication Institute at Metro North Health, Brisbane, Australia, which designs and manufactures 3D-printed medical devices at point-of-care. She is a PhD candidate at Queensland University of Technology and an adjunct research fellow at the University of Queensland.

Giorgio Andrew Katsifis (PhD) is currently employed as a medical physics registrar at Liverpool/Macarthur Cancer, NSW, Australia, where he is working towards his medical physics accreditation. His current research interests include 3D printing in medicine, radiation treatment planning for stereotactic radiotherapy and proton or heavy ion therapy. Dr Katsifis completed his PhD at the University of Sydney in 2022 and has research experience in materials, plasma, and medical physics.

Martin Leary (PhD) has been a professor of additive manufacturing with the RMIT University Centre for Additive Manufacturing (CAM), Melbourne, Australia, since its inception. Dr Leary has an expertise in experimental characterisation, numerical modelling, and topological optimisation of 3D-printed lattice structures. A focus on applied research has translated to the research and design of patient-specific radiotherapy phantoms in collaboration with the Peter MacCallum Cancer Centre.

James I Novak (PhD) is a biomedical designer who leads the CranioFacial programme at the Herston Biofabrication Institute as Senior Research Fellow. His team develops novel medical devices and research to support neurosurgical and maxillofacial applications of 3D printing, manufactured at the point of care. Dr Novak also holds an adjunct position at The University of Queensland, and is an assistant editor of the *Computer-Aided Design and Applications* journal.

Samuel C Peet (PhD) is a senior medical physicist at the Royal Brisbane and Womens Hospital and Adjunct Fellow at The University of Queensland. Dr Peet has a passion for particle therapy, software development, and the integration of new technologies in radiation oncology.

Emily Simpson-Page (MSc) is a senior medical physicist at Radiation Oncology Princess Alexandra Raymond Terrace in Brisbane, Australia. Although many facets of radiotherapy physics are intriguing to her, Emily is particularly taken by the ways in which processes can be improved at a multi-disciplinary level. Emily greatly enjoys the privilege of working in cancer care, and collaborates with colleagues to improve service delivery and patient outcomes.

Natalka Suchowerska (PhD) is an associate professor in the Department of Medical Physics, University of Sydney, Australia. A/Prof Suchowerska is an internationally recognised expert in the radiobiological effects of contemporary radiation treatment techniques with a formidable track record in the fabrication and evaluation of novel dosimetry systems. She has written over 150 publications, including six papers in 2022–2023 on the recent development of 3D-printed PEEK scaffolds for bone reconstruction.

Rance Tino (PhD) completed his doctoral research into the utility of 3D printing to develop custom radiotherapy phantoms for end-to-end testing in radiation therapy, funded through the Australian Research Council Industrial Transformation Training Centre in Additive Biomanufacturing (ARC ABM). Dr Tino is now a postdoctoral fellow at MD Anderson Cancer Center, further investigating the utility of 3D printing and computational modelling in radiation oncology.

Rachael Wilks (MSc) is the senior brachytherapy physicist at the Royal Brisbane and Women's Hospital and is currently completing her doctoral research at the University of Queensland, in Brisbane, Australia. In both her clinical and her PhD research roles, Rachael is leading the implementation of 3D-printed brachytherapy devices into clinical practice.

Adam Unjin Yeo (PhD) is the proton lead medical physicist at the Peter MacCallum Cancer Centre in Melbourne, Australia, and is affiliated with the University of Melbourne and RMIT University as an honorary research fellow. His doctoral research investigated dose delivered to deformable moving targets in radiation therapy, leading to an ongoing interest in clinical implementation of advanced imaging and new technologies. He is heavily involved in a wide range of interest groups and clinical trial committees for SRS/SABR, particle therapy, and adaptive therapy within Australasian communities.

Chapter 1

Introduction

Tanya Kairn and Tomas Kron

Modern medicine aspires towards total personalisation. Every patient is treated individually, so no one should receive a treatment that does not consider all aspects of the disease as well as the patient. This paradigm is strong in radiation medicine, where even every chest x-ray is adapted to the individual patient using collimation and automatic exposure control to optimise the balance between diagnostic value and radiation exposure.

The same is especially true for radiation therapy, where customised treatment plans and customised radiation delivery are standard practice. Although radiation therapy is an effective treatment for cancer due to the toxic effects of radiation on malignant cells, radiation is also toxic to the healthy tissues that surround the cancer. The key goal of radiation therapy is to deliver a maximum dose of radiation to the cancer, while minimising the radiation dose delivered to surrounding tissues. This goal can be achieved by carefully targeting radiation beams delivered from outside the patient via external beam radiation therapy (EBRT, as shown in figures 1.1(a)–(c)), or by placing radioactive sources inside or in contact with the patient via brachytherapy (as shown in figures 1.1(d)–(f)).

EBRT treatments are most commonly delivered using medical linear accelerators (linacs, see figure 1.1(b)), which can produce collimated electron beams to treat superficial targets, and megavoltage (MV) photon beams that can be dynamically modulated to treat deep seated tumours while sparing surrounding tissues (figure 1.1(a)). Other EBRT treatment systems range from straightforward kilo-voltage (kV) photon treatment units (figure 1.1(c)) that are specifically designed for targeting skin cancers, to the most sophisticated proton or ion beam treatment systems that can be used to minimise out-of-field dose, especially for the benefit of paediatric cancer patients.

Brachytherapy involves placing radioactive sources within or close to the treatment target, to achieve a three-dimensional (3D) dose distribution that is shaped to cover the cancer (figure 1.1(d)) using methods such as high dose rate (HDR) and low

doi:10.1088/978-0-7503-3907-0ch1

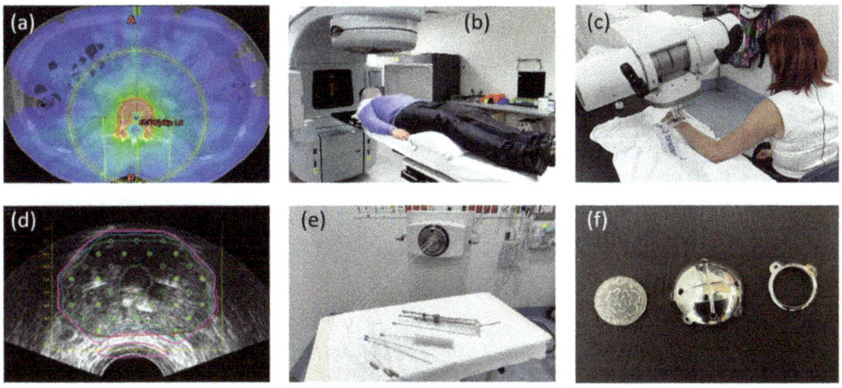

Figure 1.1. (a) EBRT treatment plan dose colourwash, showing high dose coverage of vertebral tumour and low dose to spinal cord and surrounding anatomy, (b) EBRT treatment photograph showing immobilisation mask, (c) kV photon therapy treatment photograph showing personalised lead shielding, (d) low dose rate brachytherapy treatment plan isodose display showing radioactive seeds distributed on a rectilinear grid, (e) high dose rate afterloader photograph showing generic applicators, (f) photograph of superficial (ocular melanoma) brachytherapy applicator showing plastic seed carrier.

dose rate (LDR) brachytherapy. HDR brachytherapy treatments are delivered by programming the brachytherapy afterloader to pause, or 'dwell', the source for specific dwell times at different dwell positions within an applicator or an array of catheters (figure 1.1(e)). LDR brachytherapy treatments are delivered using permanently implanted radioactive seeds (figure 1.1(d)), which gradually decay and treat the target continuously over months and years. Brachytherapy can also be delivered superficially to treat the skin or eyes (figure 1.1(f)).

These radiation therapy treatment techniques routinely use customised treatment planning and radiation dose calculations for each individual patient. This ensures that delivered dose distributions are optimised and targeted to each patient's specific anatomy in accordance with their specific treatment needs. The machines that are used to deliver radiation therapy treatments are designed with a high degree of flexibility to deliver the individually prescribed treatments as planned. Moving gantries, translating couches (sometimes called tables or beds), and dynamic collimation systems are used for MV photon treatments, a range of radiation beam-defining applicators are supplied for use in electron treatments and kV photon treatments, and a range of applicators are available for delivering brachytherapy treatments.

However, patient-specific customisation of other types of radiation therapy hardware remains both less common and more modest. Such hardware includes patient immobilisation devices, boluses, and shielding, as well as measurement jigs and patient-mimicking models called 'phantoms'. Conventional treatment devices are generic, designed for a 'standard patient' with the expectation that the individual will need to fit. While the use of standardised treatment equipment may minimise confusion and avoid opportunities for mistakes, it also limits the range of special

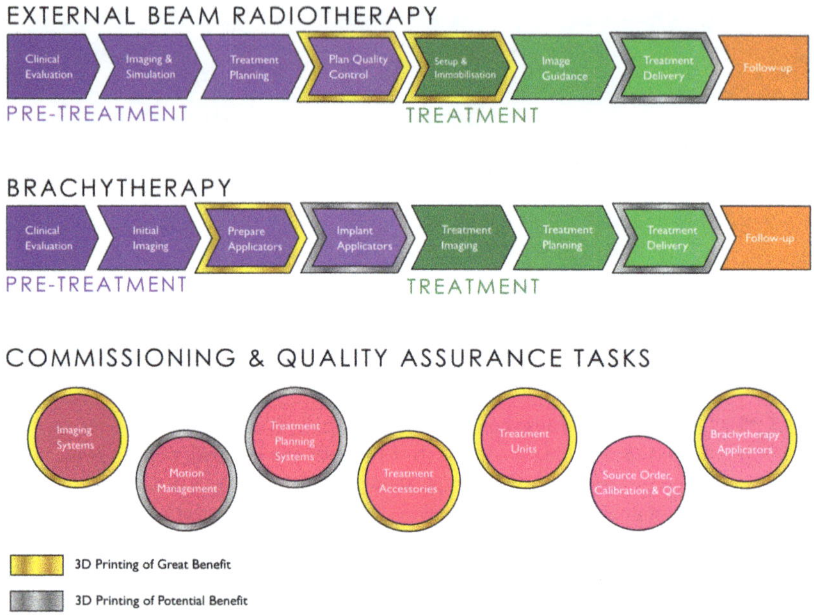

Figure 1.2. Typical patient's journey through the EBRT and brachytherapy treatment process, with opportunities for customisation via 3D printing indicated by gold and silver outlines.

procedures or one-off treatments of challenging cases that can be performed. For many radiation therapy centres, the increased flexibility derived from a programme of in-house fabrication of custom radiation therapy hardware devices would be highly advantageous.

The actions listed in the top half of figure 1.2 highlight the steps in the typical cancer patient's radiation therapy treatment process, with opportunities for hardware customisation highlighted with gold and silver outlines. Of particular importance are devices in direct contact with the patient, such as immobilisation, and beam modifying devices, such as bolus. There is a need to conform to the contours of the individual patient and at the same time be comfortable, easy to clean, and fast to make. This requirement also applies to brachytherapy where generic applicators are often used in treatments of superficial lesions, as well as within body cavities.

The tasks listed at the bottom of figure 1.2 show opportunities for the development and use of custom hardware in the commissioning and quality assurance of radiation therapy treatments. For example, the ability to perform commissioning measurements for a large variety of treatment scenarios by using realistic custom phantoms can be very helpful for optimisation and refinement of treatment techniques. Similarly, customisation can facilitate easier integration of dose measuring devices into phantoms or the inclusion of purpose-designed phantom components into existing measurement phantoms.

This book proposes additive manufacturing via 3D printing as the ideal means for radiation therapy centres to accurately and efficiently design, create, verify, and use custom devices at the point of care. Some opportunities for 3D printing to benefit the patient and improve the overall radiation therapy treatment process are shown in figure 1.2. Areas where 3D printing is already showing the greatest benefit are outlined in gold.

After first providing fundamental information regarding 3D-printing hardware, materials, software, processing, and costs (chapters 2, 3, 4, 5, and 6), this book provides specific guidance on quality management and quality assurance of radiation therapy 3D-printing services (chapters 7 and 8), before describing the growing use of 3D printing for radiation therapy patient treatments and treatment verification (chapters 9 and 10), and other medical applications (chapter 11). Specifically,

- **Chapter 2** provides details on current 3D printers, covering systems from consumer-grade to high end, and indicating the relative advantages and limitations of different printing systems for radiation therapy 3D-printing applications;
- **Chapter 3** describes the different types of materials that can be used for 3D printing, with an emphasis on plastics and plastic-composites (which are the most commonly used materials for 3D printing in radiation therapy) but also covering metals and other materials;
- **Chapter 4** describes processes for producing models for 3D printing using medical images, 3D optical surface scans, and computer-aided design systems;
- **Chapter 5** provides guidance on slicing 3D models to produce 3D-printer instructions (usually g-code), selection and effects of printing parameters, and how to start and successfully complete a 3D print, including post-processing steps;
- **Chapter 6** assesses the potential costs of 3D printing in radiotherapy (with reference to current alternatives) in terms of time and money, as well as health hazards, physical space, and waste management;
- **Chapter 7** provides guidance regarding quality management, and how a safe and effective clinical 3D-printing service can be established by adhering to a quality management framework;
- **Chapter 8** describes the quality assurance necessary for a clinical 3D-printing service by outlining the common 3D-printing defects and their consequences for radiation therapy accuracy, describing the commissioning steps needed to avoid these defects, and providing examples of maintenance and quality control steps that can be taken to detect and identify the defects;
- **Chapter 9** describes the clinical applications of 3D printing for the patient's treatment, providing guidance regarding patient safety (internal and external use, cleaning, and sterilisation), beam modifying devices (bolus, compensators, and range shifters), custom shielding (patient-specific shields, as well as electron and kV photon cutouts), patient immobilisation (mechanical safety, supports, masks, and displacement devices), and brachytherapy (superficial, interstitial, intracavitary, and dose calculation considerations);

[12] Ji Z, Jiang Y, Guo F, Sun H, Fan J, Zhang L and Wang J 2017 Dosimetry verification of radioactive seed implantation for malignant tumors assisted by 3D printing individual templates and CT guidance *Appl. Radiat. Isot.* **124** 68–74

[13] Yuan X, Zhang Y, Cui M, Miao J, Gao L, Hu J, Tian D and You J 2019 Dosimetry comparison between a 3D printed minimally invasive guidance template and free implantation in the brachytherapy treatment of postoperative recurrent cervical carcinoma *Cancer Manage. Res.* **11** 5013

[14] Wilks R, Crowe S B, Chan P, Cheuk R and Kairn T 2023 A 3D printed patient-specific vaginal mould for brachytherapy *J. Phys.: Conf. Ser.* **2630** 012026

[15] Laan R C, Nout R A, Dankelman J and van de Berg N J 2019 MRI-driven design of customised 3D printed gynaecological brachytherapy applicators with curved needle channels *3D Print. Med.* **5** 1–8

[16] Wiebe E, Easton H, Thomas G, Barbera L, D'Alimonte L and Ravi A 2015 Customized vaginal vault brachytherapy with computed tomography imaging-derived applicator prototyping *Brachytherapy* **14** 380–4

[17] Sim L 2016 Novel application of 3D printing in brachytherapy using MED610 3D printed insert for I-125 ROPES eye plaque *Australas. Phys. Eng. Sci. Med.* **39** 863–70

[18] Jones E-L, Baldion A T, Thomas C, Burrows T, Byrne N, Newton V and Aldridge S 2017 Introduction of novel 3D-printed superficial applicators for high-dose-rate skin brachytherapy *Brachytherapy* **16** 409–14

[19] Bieleda G, Chiche A, Boehlke M, Zwierzchowski G, Chyrek A, Burchardt W, Stefaniak P, Wiœniewska N, Czereba K and Malicki J 2022 3D printing of individual skin brachytherapy applicator: design, manufacturing, and early clinical results *J. Contemp. Brachyther.* **14** 205–14

[20] Arenas M, Sabater S, Sintas A, Argus M, Hernández V, Árquez M, López I, Rovirosa À and Puig D 2017 Individualized 3D scanning and printing for non-melanoma skin cancer brachytherapy: afinancial study for its integration into clinical workflow *J. Contemp. Brachyther* **9** 270–6

[21] Delaney G, Jacob S, Featherstone C and Barton M 2005 The role of radiotherapy in cancer treatment: estimating optimal utilization from a review of evidence-based clinical guidelines *Cancer* **104** 1129–37

[22] World Health Organisation 2022 *Cancer* [Online] Available: https://www.who.int/newsroom/fact-sheets/detail/cancer (visited on 05/12/2023)

IOP Publishing

3D Printing in Radiation Therapy

Tanya Kairn, Scott B Crowe and Tomas Kron

Chapter 2

3D printing

Rance Tino and Martin Leary

2.1 Introduction

Three-dimensional (3D) printing, also known as additive manufacturing (AM) or rapid prototyping, refers to a family of manufacturing technologies (figure 2.1) that fabricate 3D objects by sequentially adding input materials, either in a layer-by-layer or a volumetric approach [1, 2]. In this chapter, we explore the advantages, key components, and functions of 3D-printing technologies [3, 4]. We discuss state-of-the-art technologies used in the field of 3D printing, especially 3D printing technologies used by researchers and clinicians working in the field of radiation therapy. We focus on the commonly used 3D printers, ranging from inexpensive to high-end machines.

In recent years, 3D printing has gained enormous attention in the medical manufacturing space due its numerous advantages in comparison to the traditional manufacturing technologies of subtractive and formative manufacturing, which are designed for mass production and have limited capacity for customisation. The advantages of 3D printing for medical purposes include:

1. **Reduced costs:** 3D printing uses readily available input materials and comes at lower costs for low-volume manufacturing of customised products;
2. **Increased geometric complexity:** 3D printing uses a sequential layer-by-layer process that enables the production of complex geometries;
3. **Material customisation:** The 3D-printing process offers the possibility of developing novel materials with modified material properties;
4. **Generative design:** The prototyping nature of 3D printing allows for further design modifications with design constraints to potentially reduce costs for mass customisation;
5. **Waste reduction:** The inherent process of 3D printing allows for reduced waste and material costs; and,
6. **Distributed manufacture:** In a digital world, 3D models can be distributed globally, which enables reduced lead-time and design costs associated with transport and maintenance [3, 5].

doi:10.1088/978-0-7503-3907-0ch2

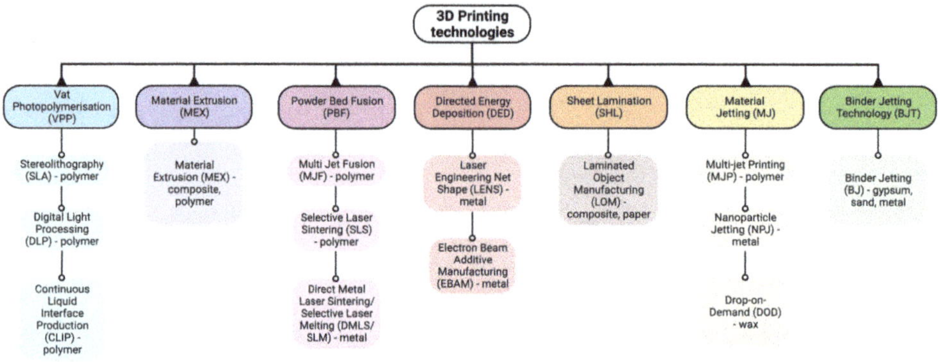

Figure 2.1. Classification of 3D-printing technologies.

The classifications of 3D-printing technologies that are summarised in this chapter include vat photopolymerisation (VPP; Section 2.2), material extrusion (Section 2.3), powder-bed fusion (Section 2.4), directed energy deposition (Section 2.5), sheet lamination (Section 2.6), material jetting (Section 2.7), and binder jetting (section 2.8). The costs, including space requirements and safety issues associated with these technologies, are addressed further in chapters 6 and 7.

2.2 Vat photopolymerisation

VPP uses the photopolymerisation process, commonly known as *curing*, where a liquid resin (usually a thermoset photopolymer resin) contained within a vat is exposed to a light of a specific wavelength, generating a solid object layer by layer, as shown in figure 2.2. Commercially available VPP systems range from inexpensive consumer-grade printers to more costly industrial systems.

The main advantages of VPP for 3D printing in radiation therapy include the ability to print at comparatively high resolutions, the potential (through careful selection of photopolymer resins and print parameters) to produce prints that are biocompatible and sterilisable (see chapter 9), and the ability to produce prints that are highly flexible (see figure 3.1(b)).

Limitations of current VPP systems include the need for support structures when printing objects with overhangs and the fact that printing of enclosed voids can be a nontrivial challenge requiring the strategic placement of drain holes, in addition to the limitation that each print job can only use one material.

VPP can be classified into different technologies based on mechanisms and light sources, as described in sections 2.2.1, 2.2.2, 2.2.3 and 2.2.4.

2.2.1 Stereolithography

Stereolithography (SLA) was the first 3D-printing technology to be developed [6, 7] and it utilises the process of curing resin liquid using an ultra-violet laser to scan or raster the desired print pattern, selectively curing the liquid resin contained in the vat,

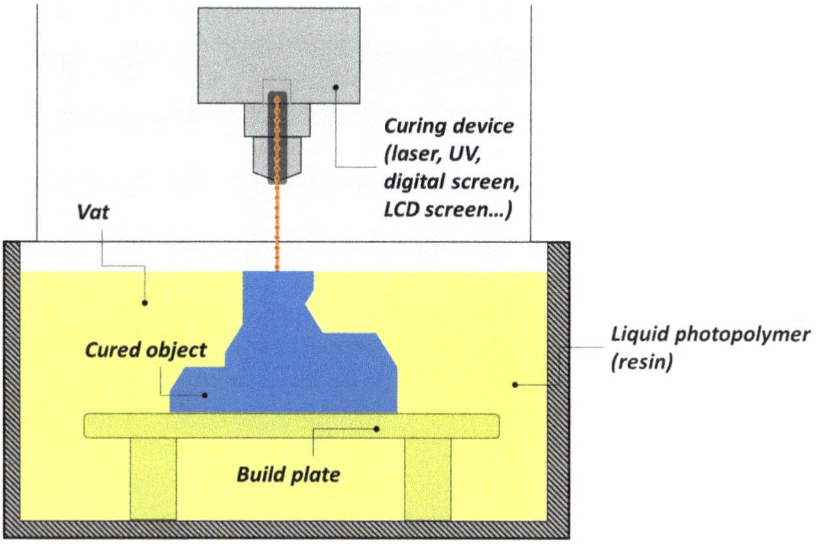

Figure 2.2. VPP process.

layer by layer. SLA can be categorised as either a 'bottom-up' printer where the ultra-violet source is positioned below the liquid resin vat, or a 'top-down' printer, where the ultra-violet source is positioned above the build platform and the vat.

SLA printers are becoming more useful in the medical field as they enable the rapid fabrication of high-resolution devices. Care must be taken when designing objects with internal voids for printing using SLA systems because the 'bottom-up' and 'top-down' methods require different approaches to the positioning of drain holes. Note that drain holes often need to be placed on both the upper and lower surfaces of the internal void, regardless of the direction of printing, so that a vacuum effect does not prevent draining.

2.2.2 Digital light processing

Digital light processing (DLP) uses a similar method to SLA for producing solid objects. The main difference between DLP and SLA is that DLP uses a digital light projector screen rather than an ultra-violet laser to cure the liquid resin contained in the vat [8–10].

DLP has a faster printing time compared to SLA because each entire layer is exposed simultaneously rather than having to selectively outline the cross-sectional area with an ultra-violet laser. Conventional DLP printers cover a high printing resolution, ranging from 0.6 to 90 microns, by modulating the magnification of the projection lens.

2.2.3 Continuous direct light processing

Continuous direct light processing (CDLP), which is also known as continuous liquid interface production (CLIP) [11], produces solid objects using the same photopolymer exposure method as DLP. However, CDLP uses a continuous upward motion of the build platform rather than stopping the build platform at each successive layer, which allows for a faster printing time compared to SLA and DLP. CDLP can create objects from a wide range of materials, including polymers and composites with ceramics and metals, via material doping techniques. CDLP can produce printed objects with higher resolution and more complex geometries than SLA and DLP.

One of the key features of CDLP is that it allows for the creation of objects with properties that are difficult or impossible to achieve with traditional manufacturing methods. For example, CDLP can be used to create objects with complex internal structures or objects with graded material properties. CDLP has the potential to revolutionise the way in which products are manufactured and it is already being used in a wide variety of applications, including aerospace, automotive, and consumer products [12].

2.2.4 State-of-the-art vat photopolymerisation-based techniques

Although they have limited printable volumes when compared to other 3D-printing technologies, SLA, DLP, and CDLP technologies provide printing resolutions ranging from 25 to 300 microns, which is ideal for users who require high-resolution surface finishing of printed objects at relatively fast printing speeds. Additionally, several recent developments have further refined the VPP process.

Projection micro-stereolithography (PμSL) [9, 13–18] is a DLP-based 3D-printing technique that uses a digital micromirror device as a dynamic mask to produce higher 3D-printing resolutions, ranging from 0.6 to 30 microns, with larger build volumes than previously achievable.

Some studies have started to explore the possibilities of sub-micron 3D-printing resolutions using two-photon polymerisation (TPP), which is also known as two-photon lithography (2PL), to control the pulse energy and the number of applied pulses of the laser source, which effectively cures photopolymers at ultra-high resolutions of 100 nm or better [19–22].

More recently, volumetric-based stereolithography 3D-printing techniques [23, 24] have emerged. These techniques apply the principles of computed tomography (CT), where multiple images are projected into the target photopolymer volume at defined angles to rapidly cure the entire 3D object to achieve a faster print time than TPP.

Finally, xolography [25] makes use of dual-colour photopolymerisation (DCP) process [26] to achieve better resolutions than volumetric-based stereolithography techniques with faster volume generation rates.

2.3 Material extrusion (also known as FDM or FFF)

Material extrusion (MEX), commonly known as fused deposition modelling (FDM) or fused filament fabrication (FFF), was made famous by the Replicating Rapid Prototyper project [27]. In MEX, polymers or composites, in the form of a filament or pellets, are melted and extruded onto a build plate. The nozzle moves along a

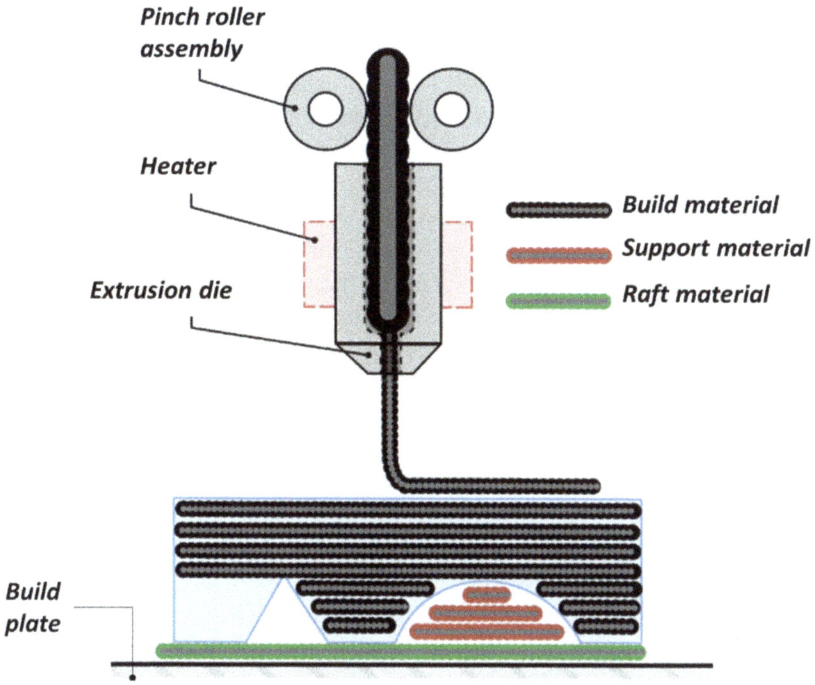

Figure 2.3. MEX (aka FDM or FFF) process. Reprinted from [5], Copyright (2019), with permission from Elsevier.

predetermined path and deposits layers of material in a specific pattern to create the desired object, as shown in figure 2.3.

Like VPP, MEX prints may need support structures for overhangs. Although MEX prints generally have a lower resolution than VPP, MEX printers for radiation therapy use can include multiple materials in one print [28, 29]. MEX printers also have greater accessibility and are more affordable than the other printing systems described in this chapter.

MEX has been reported as the most commonly used 3D-printing technology in radiation therapy centres and radiation therapy research [2, 30, 31], which is unsurprising given that MEX printers are inexpensive, reliable, and usable by both entry-level and professional users. MEX printers in radiotherapy centres are often handled by biomedical engineers, medical physicists, mould room technicians, and radiotherapists, who work closely with research students and in collaboration with radiation oncologists, to develop tools and devices for research and clinical use, specifically for patient treatments (see chapter 9), and for treatment machine quality assurance and radiotherapy treatment verification (see chapter 8).

Differences between MEX printer models include the extent of support for different materials (maximum temperature), print volume (build plate size and height), extrusion type (direct versus Bowden tube), build plate type (glass, polyetherimide (PEI) sheet), and the use of an open frame or enclosure.

The selection of a MEX printer for radiation therapy device printing will depend on the types and range of devices to be printed, and the required properties of those devices (e.g. size, flexibility, and sterility). Supported materials will be determined by the maximum temperature of the hot end system and the available nozzle diameters. For phantoms or large devices, a large print volume (in terms of both build plate size and height) will be required. A printer that is able to print quasi-simultaneously with multiple materials (either by multiple nozzles or an automatic filament substitution system) will allow greater flexibility when printing phantoms or models with multiple colours or densities. When printing flexible materials, a direct extrusion system (where the gears that extrude the filament are directly above the hot-end and nozzle) may perform better than a system with a Bowden tube, although heat creep can be an issue for long prints.

Entry-level MEX printers that use open-source designs or slicing software can often be upgraded and their components can easily be replaced. The downside of these entry-level products is that more effort is required to optimise the printer's settings, and maintain correct calibration and build plate levelling (see chapter 8). In contrast, mid-to-high level MEX systems, such as the Raise3D systems, are easy to use but rely on proprietary software.

Common MEX printer upgrades include nozzle substitution, replacement of the build plate (e.g., with glass or a flexible sheet), and the use of all-metal hot ends. Common replacements include the plate surface (e.g., due to damage to the PEI sheet), the nozzle (due to wear or abrasion from filaments containing metals or other particulates), and small electrical components such as fans and thermistors.

The concept of MEX 3D-printing has evolved towards complex biofabrication systems involving the bioprinting of hydrogels or bioinks combined with cells for the fabrication of tissue scaffolds or resorbable implants (see chapter 11). Direct-ink-writing (DIW) or electrohydrodynamic (EHD) printing of liquid-phase materials, has expanded this technology to a wide range of printable materials. Melt-electro-writing (MEW) is a recently developed biofabrication technique that combines the principles of melt electrospinning and material extrusion technology, where materials are electrically charged and extruded under a high-voltage field, leading to the solidification of the materials. Here, MEW can fabricate objects with improved z-resolutions, as low as 0.8 microns [32].

2.4 Powder bed fusion

Powder bed fusion (PBF) utilises a thermal source that leads to the sintering or melting of polymer, ceramic, or metal powder particles on a platform (bed) to produce a solid object, layer by layer, as shown in figure 2.4. The major advantage of PBF over most other 3D-printing processes is that PBF can be used to 3D print metal objects [33–35]. PBF printing also conserves printing material by eliminating the need for overhang supports and by allowing all non-melted material in the bed to be re-used (as discussed in chapter 6).

The key limitations of PBF systems, in terms of their potential use in clinical environments, relate to the physical size and relative expense of the printers, the

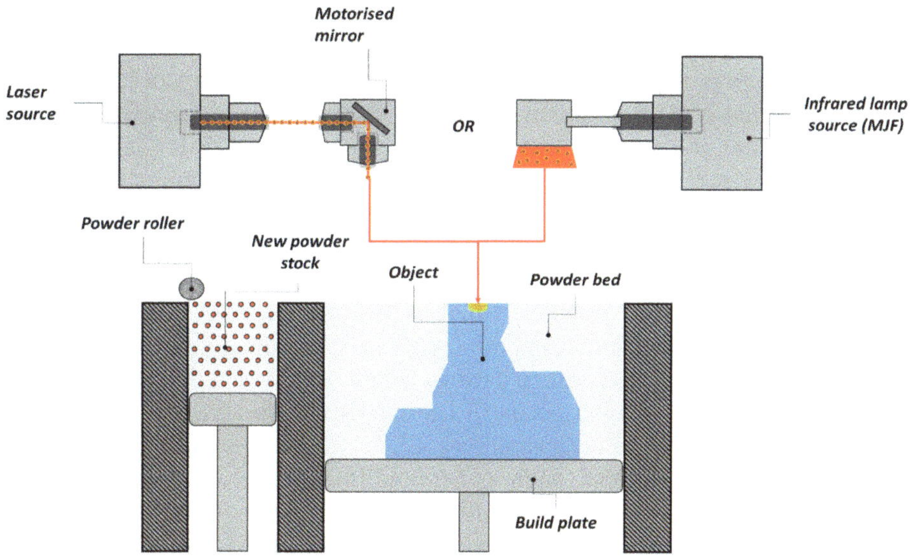

Figure 2.4. PBF process.

requirement for air-tight enclosures or explosion-proof shielding (including built-in shielded housings) for some BPF systems, and the inhalation risks posed by the powders that they use. An additional limitation relating to PBF use in radiation therapy phantom production is that only one material can be used at a time. Changing PBF printing materials can take hours to days, depending on the specific type of printer used.

PBF can be classified into four technologies based upon the energy sources and the types of powder used in the manufacturing process, as described in sections 2.4.1, 2.4.2, 2.4.3 and 2.4.4.

2.4.1 Multi jet fusion

Multi-jet fusion (MJF) technology was pioneered by Hewlett-Packard. It uses an inkjet array to selectively apply fusing and detailing agents across a bed of nylon powder, which are then fused (using an infra-red lamp) into a solid layer. After each layer, powder is distributed on top of the bed using a powder roller and the process repeats until the object is complete. When the build finishes, the entire powder bed with the encapsulated objects is moved to a processing station where most of the loose powder is removed by an integrated vacuum. The printed objects are then bead blasted to remove any of the remaining residual powder before ultimately reaching the finishing compartment where they can be dyed to improve their appearance.

2.4.2 Selective laser sintering

Selective laser sintering (SLS) uses a laser to fuse powdered polymer, composite, ceramics, or metals together to create solid objects. The process begins by laying down a thin layer of powder on a build platform. The laser then scans the outline of the desired object onto the powder, sintering or melting it together to form a solid layer. This process is repeated layer by layer until the entire object has been built. One of the benefits of SLS is that it allows for the creation of complex shapes and geometries because the laser can easily access and sinter powder in any location within the build space. SLS is also a relatively fast process, with many machines able to print at speeds of up to 100 mm per second. It is often used in the production of prototypes and low-volume manufacturing applications, as well as in the creation of functional end-use parts. The most common material powder utilised for SLS is an engineering-grade Nylon 12 powder, which is also known as PA12.

In healthcare, SLS is often used to create custom drug delivery devices and prosthetics. For example, SLS can be used to create customised drug delivery devices that are tailored to the specific needs of an individual patient with capabilities to modify the rate at which the drug dissolves. SLS can also be used to create patient-specific prosthetics or orthotics, such as personalised arm braces or surgical guides (see chapter 11). SLS allows for the creation of complex geometries and fine details. SLS systems can also print with a variety of different materials, including metals and polymers. SLS is relatively fast and, beyond the substantial cost of acquiring the printer, is a cost-effective manufacturing method, which makes it an attractive option for the healthcare industry.

2.4.3 Direct metal laser sintering and selective laser melting

Direct metal laser sintering (DMLS) and selective laser melting (SLM) use lasers to fuse powdered metal alloys to create solid metal objects [35]. Similar to the SLS process, both DMLS and SLM work by laying down a thin layer of powder on a build platform and then using a laser to scan the outline of the desired object onto the powder, sintering or melting it together to form a solid layer. This process is repeated layer by layer until the entire object has been built.

The main difference between DMLS and SLM is the level of melting that occurs. DMLS uses a laser to sinter the powdered material together by heat and pressure, which means that the particles are melted enough to fuse together but the material remains in a solid state. Meanwhile, SLM uses a laser to melt the powdered material together, resulting in a fully molten pool of material, with particles that combine and solidify as they cool down. Additionally, SLM uses single metals whereas DMLS can use metal alloys [33]. To account for this, in SLM only one melting or fusing temperature is required for bonding, whereas in DMLS several are needed [33].

One advantage of SLM is that it allows for the production of objects with a high level of structural integrity and density because the material is fully melted and solidified during the process. This makes it an ideal choice for the production of high-strength metal parts. DMLS is also a strong process. However, it produces prints with slightly lower density and strength due to the sintering process. DMLS is often used to produce objects with high accuracy and fine detail because the metal powders used in the process have a fine particle size and can be sintered with high precision.

2.4.4 Electron beam melting

Electron beam melting (EBM) uses an electron beam to melt and deposit layers of metal powder to create an object. Similar to the laser sintering and melting processes, EBM uses a layer of metal powder that is placed onto the build platform and an electron beam is directed onto the powder to melt it. The melted material is then solidified, and the process is repeated with successive layers of powder until the object is complete.

EBM is primarily used to create metal parts, though the types of metals that can be used in the process are limited. Materials suitable for the EBM process include titanium and chromium-cobalt. Objects manufactured via EBM are limited in size because the printers have relatively small build chambers, so EBM may not be suitable for creating very large prints. Furthermore, EBM's accuracy is lower when compared to laser sintering and melting technologies due to slightly wider electron beams.

2.5 Directed energy deposition

Directed energy deposition (DED) generates printed objects by melting powder material as it is ejected out of the nozzle, as shown in figure 2.5. DED printing systems are generally large and comparatively expensive. While print resolution is limited, the advantage of DED systems (like other PBF systems) is their ability to fabricate objects from a wide range of materials, including metal. DED primarily uses metallic powder or wire as an input material for producing 3D-printed metal objects [33–35].

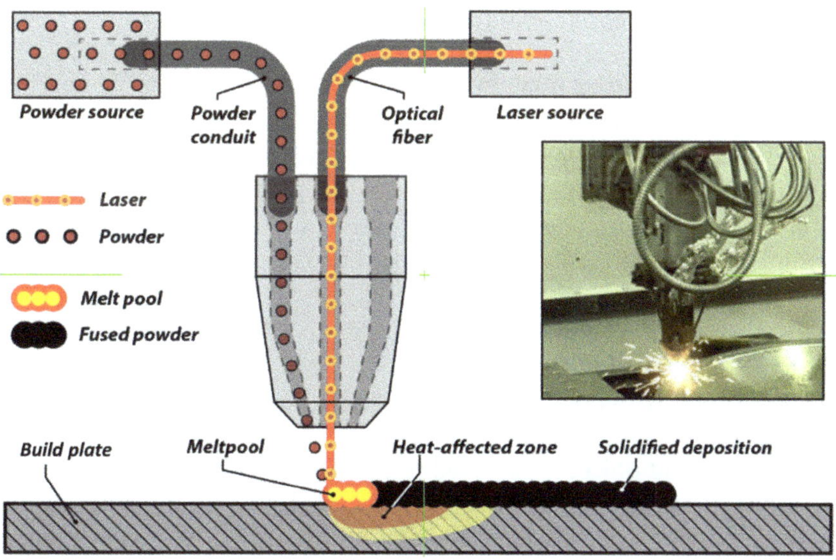

Figure 2.5. DED process. Reprinted from [5], Copyright (2019), with permission from Elsevier.

In DED methods, a nozzle deposits the metal onto a surface and it is fused using radiation. During the printing process, the metal structures are usually assembled with supports, which are made from the same metal, to mitigate any deformation of the printed object at the high temperatures during printing. When the chamber cools to room temperature, the excess powder is manually removed and the prints are heat treated while still attached to the build platform to relieve any residual stresses. The components are then detached from the build plate via cutting, machining, or wire EBM, and are ready for use or further post-processing [33].

DED has received significant attention in recent years because of its potential for use in diverse material processing activities, such as metallic coating, high-value component repair, prototyping, and even low-volume manufacturing. DED can be further classified by its two heat sources, lasers or electron beams, which are described in sections 2.5.1 and 2.5.2, respectively.

2.5.1 Laser engineering net shape

Laser engineering net shape (LENS) uses a laser source to melt the ejected powder from the dispensing nozzle to build a solid object layer by layer. Here, LENS can manufacture materials including powdered metals, alloys, ceramics, or composites. The LENS process must take place in a sealed chamber that is filled with argon to ensure that the oxygen and moisture levels stay very low. The powder material is directly delivered to the nozzle. Once a single layer has been deposited, the nozzle moves on to the next layer. After printing, the component is removed and can then be heat treated, hot isostatic pressed, machined, or finished in any required fashion.

Due to its large nozzle size, LENS can often produce uneven surfaces and requires subtractive manufacturing processes such as laser cutting or machining to achieve the desired print geometry. Thus, models planned for LENS 3D printing need to be designed with larger margins to compensate for this limitation.

2.5.2 Electron beam additive manufacturing

Similar to EBM, electron beam additive manufacturing (EBAM) utilises electron beam sources to melt the extruded material. However, EBAM uses a wire feedstock system, where wires are fed through the nozzle and are then melted using the electron beam heat source. The chamber is evacuated for EBM [35]. This process is currently limited to one manufacturer (Sciaky Inc, Chicago, USA), which may limit the availability of this technology.

2.6 Sheet lamination

Sheet lamination (SHL) is the process of manufacturing a solid object by stacking and laminating thin sheets of paper, polymer, or metal, as shown in figure 2.6. The final shape of the solid object is achieved using subtractive manufacturing by means of laser cutting or machining. SHL can be further classified into seven technologies that are varied in terms of the materials used and the lamination method, although the most relevant SHL technology for use in clinical radiation therapy 3D printing [36] (and the one discussed here) is laminated object manufacturing (LOM).

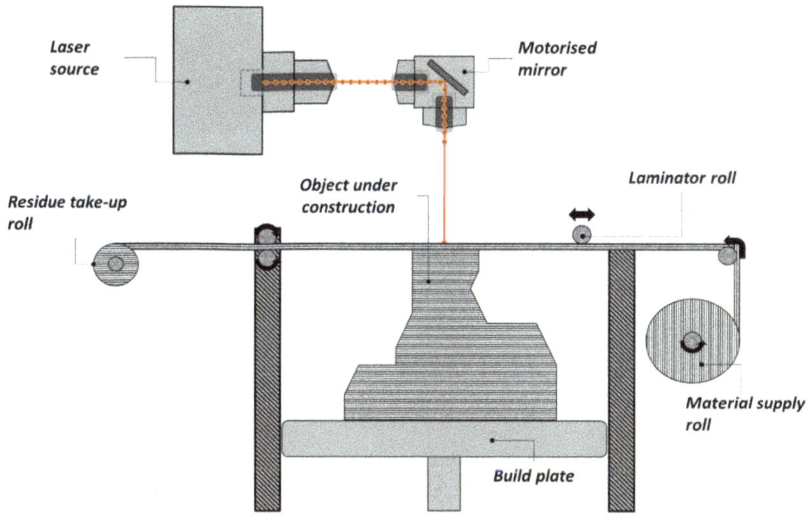

Figure 2.6. The SHL process.

In the LOM process, sheets of material are placed on a build platform and coated with an adhesive layer. The adhesive is heated by a feeding roller to melt it and bond each layer to the previous one. A blade or laser is used to shape the object and remove excess material. The process is repeated by moving the build platform down and adding another layer of material until the model or prototype is complete. If the object is made using layers of paper, it may have wood-like properties and require sanding or sealing with paint or lacquer to protect against moisture. LOM is frequently used for rapid prototyping because it is a relatively quick and (beyond the cost of the printer) affordable process.

The level of accuracy in the final object depends on the thickness of the material layers used. LOM is usually not as precise as other SHL technologies, such as selective lamination composite object manufacturing (SLCOM), plastic sheet lamination (PSL), computer-aided manufacturing of laminated engineering materials (CAM-LEM), selective deposition lamination (SDL), composite based additive manufacturing (CBAM), or ultrasonic additive manufacturing (UAM).

2.7 Material jetting

Material jetting (MJ) can be considered to be a 3D analogue of the commonly known 2D inkjet printing process. MJ uses the photopolymerisation process of curing thermoset polymers via ultra-violet light to produce a solid object, layer by layer, as shown in figure 2.7.

MJ printers can be very expensive when compared to consumer-grade VPP or MEX systems and overhangs can be printed using removable support materials. Although it is possible to fabricate hollow objects using MJ, do keep in mind that they will automatically be printed with a support material inside the hollow space to ensure that the external surface is printed properly. This can be resolved by

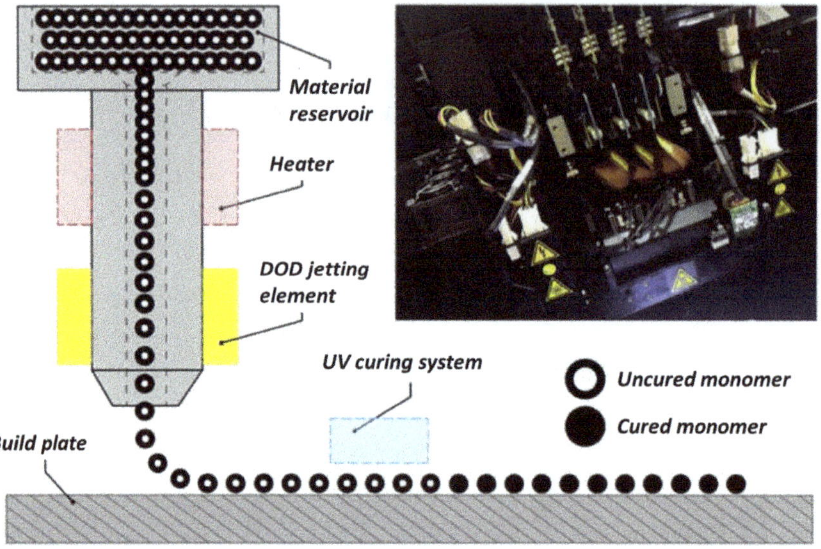

Figure 2.7. MJ proces. Reprinted from [5], Copyright (2019), with permission from Elsevier.

designing a reasonably sized hole on the hollowed object for the removal of the support material or the hollow objects may be designed and printed in halves with connectors for assembly after printing. The advantage of MJ over other 3D-printing technologies is its multi-material printing capability, which can use three or more materials.

MJ can be classified into three technologies based on the materials and the extrusion process used, as described in sections 2.7.1, 2.7.2, and 2.7.3.

2.7.1 Polymer multi-jet printing

Polymer jetting is the most common form of multi-jet printing (MJP) and involves the deposition of droplets of photopolymer resin onto a build platform to create a 3D object. In a polymer jetting printer, the liquid resin is held in a container (called an inkjet printhead), which contains nozzles that are activated by piezoelectric or thermal actuators to release precise amounts of material onto the build platform. The material is then cured or solidified using an ultra-violet light or heat source.

Polymer jetting technology can produce high resolution and fine detail in 3D-printed objects. Its ability to produce smooth surfaces and sharp edges make polymer jetting suitable for applications that require precision, such as the production of dental crowns and jewellery. This technology also has the advantage of being able to use multiple materials in a single print. Some printers have multiple printheads that can be loaded with different materials, enabling the creation of objects with multiple colours or different material properties.

The limitations of polymer jetting include slower print speeds than VPP and a limited range of materials when compared to technologies such as MEX or SLS.

Advances in printhead technology and material development are increasing the capabilities of polymer jetting printers, which makes them a valuable tool for industries and applications that require high resolution and the ability to use multiple materials in a single build.

2.7.2 Nanoparticle jetting

Nanoparticle jetting (NPJ) is a variation of polymer jetting that uses photopolymer resins doped with nanoparticles. One of the main advantages of NPJ is its ability to produce high resolution and fine detail in 3D-printed objects. The small size of the nanoparticles allows for a higher level of precision and smoothness in the final printed object. This makes nanoparticle jetting suitable for a wide range of applications, including microelectromechanical systems, medical devices, and precision parts. Another benefit of nanoparticle jetting is its versatility in terms of material options. The resin can be formulated with a wide range of nanoparticles, including metals, ceramics, and polymers, allowing for the creation of objects with different material properties.

There are a few limitations to NPJ technology: NPJ is generally slower than other 3D-printing methods; the material options are limited when compared to technologies such as MEX or SLS; and, like other forms of MJP, NPJ can only fabricate objects with overhangs and hollow centres if they are designed and printed to allow the removal of support materials. In addition, the cost of the equipment and materials can be relatively high.

Overall, NPJ is a valuable tool for industries and applications that require high resolution and fine detail. Its versatility in terms of material options, and ability to produce smooth and precise objects, make it an attractive option for a wide range of applications.

2.7.3 Drop-on-demand

Drop-on-demand (DOD) technology uses the same concept of droplet deposition of photopolymer resins as other MJP technologies. DOD technology is often used to create models for lost-wax casting or investment casting and mould making applications, making it an indirect 3D-printing method.

2.8 Binder jetting

Binder jetting (BJT) selectively dispenses droplets of binding agents onto a bed of powder (metal or polymer) to produce a solid object, layer by layer, as shown in figure 2.8. BJT is similar to inkjet printing but instead of ink BJT uses a binder fluid to bond the powdered material together. BJT process starts by spreading the powdered material over the build platform. The binder fluid is then jetted through a printhead onto the powder to bond the particles together. The build platform is lowered, and the process is repeated until the object is complete.

The BJT method is efficient and fast, allowing for the rapid production of large objects at a lower print cost (beyond the substantial up-front cost of the printer) due to its minimal material usage. BJT can also use a variety of materials (including

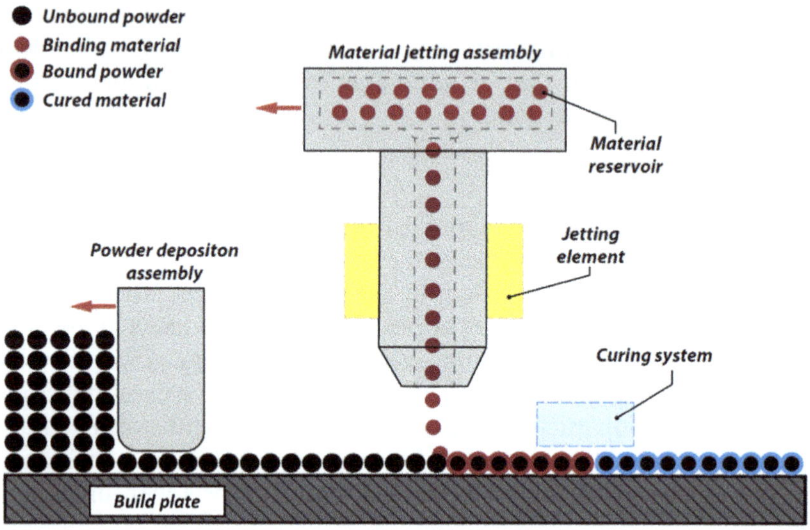

Figure 2.8. BJT process. Reprinted from [5], Copyright (2019), with permission from Elsevier.

metals, sand, and ceramics) and can produce objects with multiple colours or properties through the use of different binders or powder mixtures. However, BJT may not be as precise as other 3D-printing techniques, and the finished product may have lower resolution and less smooth surfaces. In addition, the binder fluid can be difficult to remove, requiring additional post-processing.

2.9 Key points

- 3D printing, which is also known as AM or rapid prototyping, is the process of creating 3D physical objects in a layer-by-layer or volumetric process.
- 3D printing can be achieved using a wide variety of hardware systems, which can be broadly categorised into seven types: VPP, MEX, PBF, DED, SHL, MJP, and BJT (defined in the list of acronyms and initialisms at the end of the book).
- MEX (also known as FDM or FFF) is increasingly being adopted as a tool for radiation therapy device fabrication (see chapters 9 and 10), due to low cost, ease of use, and upgrade pathways.
- When purchasing a MEX printer, commonly desired features include a large print volume, support for multiple materials in a single print, support for flexible materials, and direct extrusion.
- The use of SLS for radiation therapy applications is growing thanks to the increased precision that can be achieved using this method, as well as its ability to 3D print metal objects.
- VPP printers are also useful for radiation therapy applications because of their ability to produce high-resolution prints from potentially biocompatible and sterilisable materials (see chapter 9).

- The implementation of 3D-printing technologies in radiation therapy centres will vary depending on the clinical needs of each centre, and in particular the required properties of the devices to be printed.
- Cost considerations, safety, and resourcing are of the utmost importance (see chapters 6 and 7 for more information).

References

[1] Asfia A, Novak J I, Mohammed M I, Rolfe B and Kron T 2020 A review of 3D printed patient specific immobilisation devices in radiotherapy *Phys. Imaging Radiat. Oncol.* **13** 30–5

[2] Tino R, Leary M, Yeo A, Kyriakou E, Kron T and Brandt M 2020 Additive manufacturing in radiation oncology: a review of clinical practice, emerging trends and research opportunities *Int. J. Extreme Manuf.* **2** 012003

[3] Gibson I, Rosen D, Stucker B, Khorasani M, Rosen D, Stucker B and Khorasani M 2021 *Additive Manufacturing Technologies* (Berlin: Springer)

[4] International Organization for Standardization 2015 *ISO 17 296-2:2015 Additive Manufacturing—General Principles—Part 2: Overview of Process Categories and Feedstock* 1st edn (Geneva: ISO)

[5] Leary M 2019 *Design for Additive Manufacturing* (Amsterdam: Elsevier)

[6] Hull C W 1984 Apparatus for production of three-dimensional objects by stereolithography US Patent US4575330A

[7] Jacobs P F 1992 *Rapid Prototyping & Manufacturing: Fundamentals of Stereolithography* (Southfield, MI: Society of Manufacturing Engineers)

[8] Lu Y, Mapili G, Suhali G, Chen S and Roy K 2006 A digital micro-mirror device-based system for the microfabrication of complex, spatially patterned tissue engineering scaffolds *J. Biomed. Mater. Res.* A **77** 396–405

[9] Han L-H, Mapili G, Chen S and Roy K 2008 Projection microfabrication of three-dimensional scaffolds for tissue engineering *J. Manuf. Sci. Eng.* **130** 021005

[10] Choi J-W, Wicker R, Lee S-H, Choi K-H, Ha C-S and Chung I 2009 Fabrication of 3D biocompatible/biodegradable micro-scaffolds using dynamic mask projection microstereolithography *J. Mater. Process. Technol.* **209** 5494–503

[11] Tumbleston J R *et al* 2015 Continuous liquid interface production of 3D objects *Science* **347** 1349–52

[12] Poelma J and Rolland J 2017 Rethinking digital manufacturing with polymers *Science* **358** 1384–5

[13] Sun C, Fang N, Wu D and Zhang X 2005 Projection micro-stereolithography using digital micro-mirror dynamic mask *Sensors Actuators* A **121** 113–20

[14] Zheng X *et al* 2014 Ultralight, ultrastiff mechanical metamaterials *Science* **344** 1373–7

[15] Ge Q, Sakhaei A H, Lee H, Dunn C K, Fang N X and Dunn M L 2016 Multimaterial 4D printing with tailorable shape memory polymers *Sci. Rep.* **6** 1–11

[16] Zheng X *et al* 2016 Multiscale metallic metamaterials *Nat. Mater.* **15** 1100–6

[17] Kowsari K, Akbari S, Wang D, Fang N X and Ge Q 2018 High-efficiency high-resolution multimaterial fabrication for digital light processing-based three-dimensional printing *3D Print. Addit. Manuf.* **5** 185–93

[18] Ge Q, Li Z, Wang Z, Kowsari K, Zhang W, He X, Zhou J and Fang N X 2020 Projection micro stereolithography based 3D printing and its applications *Int. J. Extreme Manuf.* **2** 022004

[19] Kawata S, Sun H-B, Tanaka T and Takada K 2001 Finer features for functional microdevices *Nature* **412** 697–8

[20] Sun H-B and Kawata S 2004 Two-photon photopolymerization and 3D lithographic microfabrication *NMR 3D Analysis Photopolymerization* (Berlin: Springer) pp 169–273

[21] Wu S, Serbin J and Gu M 2006 Two-photon polymerisation for three-dimensional microfabrication *J. Photochem. Photobiol.* A **181** 1–11

[22] Saha S K, Wang D, Nguyen V H, Chang Y, Oakdale J S and Chen S-C 2019 Scalable submicrometer additive manufacturing *Science* **366** 105–9

[23] Kelly B E, Bhattacharya I, Heidari H, Shusteff M, Spadaccini C M and Taylor H K 2019 Volumetric additive manufacturing via tomographic reconstruction *Science* **363** 1075–9

[24] Loterie D, Delrot P and Moser C 2020 High-resolution tomographic volumetric additive manufacturing *Nat. Commun.* **11** 852

[25] Regehly M, Garmshausen Y, Reuter M, König N F, Israel E, Kelly D P, Chou C-Y, Koch K, Asfari B and Hecht S 2020 Xolography for linear volumetric 3D printing *Nature* **588** 620–4

[26] Swainson W K 1971 Method, medium and apparatus for producing three-dimensional figure product US Patent US4041476A

[27] Jones R, Haufe P, Sells E, Iravani P, Olliver V, Palmer C and Bowyer A 2011 RepRap–the replicating rapid prototyper *Robotica* **29** 177–91

[28] Tino R B, Yeo A U, Brandt M, Leary M and Kron T 2022 A customizable anthropomorphic phantom for dosimetric verification of 3D-printed lung, tissue, and bone density materials *Med. Phys.* **49** 52–69

[29] Kairn T, Zahrani M, Cassim N, Livingstone A, Charles P and Crowe S 2020 Quasi-simultaneous 3D printing of muscle-, lung-and bone-equivalent media: a proof-of-concept study *Phys. Eng. Sci. Med.* **43** 701–10

[30] Albantow C E and Brown S J 2022 Three-dimensional printer use in the Australian and New Zealand radiation therapy setting *J. Med. Radiat. Sci.* **69** 492–501

[31] Rooney M K, Rosenberg D M, Braunstein S, Cunha A, Damato A L, Ehler E, Pawlicki T, Robar J, Tatebe K and Golden D W 2020 Three-dimensional printing in radiation oncology: a systematic review of the literature *J. Appl. Clin. Med. Phys.* **21** 15–26

[32] Loewner S, Heene S, Baroth T, Heymann H, Cholewa F, Blume H and Blume C 2022 Recent advances in melt electro writing for tissue engineering for 3D printing of microporous scaffolds for tissue engineering *Frontiers Bioeng. Biotechnol.* **10** 896719

[33] Duda T and Raghavan L V 2016 3D metal printing technology *IFAC-PapersOnLine* **49** 103–10

[34] Sing S, Tey C, Tan J, Huang S and Yeong W Y 2020 3D printing of metals in rapid prototyping of biomaterials: techniques in additive manufacturing *Rapid Prototyping of Biomaterials* (Amsterdam: Elsevier) pp 17–40

[35] Pidge P A and Kumar H 2020 2020 Additive manufacturing: a review on 3D printing of metals and study of residual stress, buckling load capacity of strut members *Mater. Today: Proc.* **21** 1689–94

[36] Tino R, Yeo A, Leary M, Brandt M and Kron T 2019 A systematic review on 3D-printed imaging and dosimetry phantoms in radiation therapy *Technol. Cancer Res. Treat.* **18** 1533033819870208

Chapter 3

Materials

Amirhossein Asfia, Giorgio Andrew Katsifis, James I Novak and Scott B Crowe

3.1 Introduction

While the chosen three-dimensional (3D) printing hardware (described in chapter 2), combined with the digital input (see chapters 4 and 5), plays a critical role in the functional and aesthetic properties of the resulting object or device, the material used for 3D printing also contributes significantly to the quality of the object. This is especially important within a medical context, where issues such as biocompatibility, tolerance to sterilisation, and dosimetric properties are key considerations, alongside more functional concerns such as strength, flexibility and appearance [1]. This chapter provides an overview of the key properties and dosimetric characteristics of the 3D-print materials that are used for radiation therapy applications, including plastics and metals. Biocompatibility and sterilisation are addressed generally in chapter 6 and more specifically in relation to patient treatment applications in chapter 9.

Polymers (commonly known as plastics) are a low-cost and diverse family of materials that have found value in a broad range of applications, from consumer electronics to packaging and textiles [2]. They were also the first type of material to be utilised for 3D printing [3], initially as a photocurable resin during the 1980s, before extending to powders, filaments, and others as the technologies of 3D printing evolved. In general, plastics can be categorised as being either thermoset or thermoplastic. Thermoset plastics, such as the photocurable resins used in vat photopolymerisation 3D printing, experience an irreversible solidification process upon curing, meaning that they cannot be reheated and reused like thermoplastics [4]. This is due to the complex chemical links formed between plastic molecules during curing and can result in printed objects with good resistance to solvents and other chemicals compared to thermoplastics. Meanwhile, thermoplastics soften into a liquid-like state when heated to their glass transition temperature or melting temperature, and once cooled into a solid can be repeatedly reheated and re-formed. Although repeated heating and cooling will affect the material characteristics,

doi:10.1088/978-0-7503-3907-0ch3

additives or fresh material may be mixed with the thermoplastic to maintain desirable performance. Whereas thermoset plastics are typically in a liquid resin form before curing, thermoplastics come in numerous forms for 3D printing, including filaments, powders, and pellets. While the sustainability of plastics is an important issue, particularly through their growing use within 3D-printing technologies [5], within the medical context they offer important qualities that are unmatched by other materials. These will be explored in the following sections, which focus on the use of plastics for 3D printing in radiation therapy, in both pure polymer form (section 3.2) and in composite materials (section 3.3).

The evolution of 3D printing from a rapid prototyping tool to commercial manufacturing has also seen the introduction of metallic 3D printing, particularly in the aeronautical and medical industries. These methods have allowed the production of objects with complex shapes and internal architectures that are difficult to obtain using traditional manufacturing methods such as casting or powder metallurgy [6]. Metal objects have played and continue to play a vital role in radiation therapy, either in the construction of equipment or the construction of beam modifiers such as wedges or collimators. They can also be used as bolus and shields to control the dose distribution to the patient. Denser metals such as lead and tungsten have been used as radiation shielding, due to their high atomic number and density. 3D printing with metals is discussed in section 3.4.

The final section of this chapter addresses other materials that can be used in radiation therapy applications through the use of 3D-printed moulds (section 3.5)

3.2 3D-printed plastics

A wide variety of plastic materials can be used for 3D printing of radiation therapy devices—some examples are shown in figure 3.1. Asfia *et al* [8] conducted a review of 3D-printing techniques to fabricate immobilisation devices, and found that acrylonitrile butadiene styrene (ABS) and polylactic acid (PLA) have been the most used materials to date. This aligns with the finding that material extrusion (MEX, aka FDM or FFF) was the most used print technology for manufacturing these objects, as outlined in chapter 2. Tino *et al* [1] also investigated the 3D-printing techniques that are used for manufacturing radiation therapy tools, and found that 3D printing is increasingly used to create high-resolution and multi-material tools, such as phantoms, boluses, immobilisation devices, brachytherapy moulds, and compensators. The range of 3D-printing technologies and their associated plastic materials are summarised in table 3.1. These are some of the more accessible and commercially-available plastic 3D-printing materials, at the time of writing. However, material development for 3D printing is a fast-paced field and new variations of materials are frequently developed, and therefore frequent literature searches should be used to identify the best material for each 3D-printing purpose.

ABS has dimensional stability, resistance to environmental and chemical attack, high toughness, and good thermal stability and rigidity [9]. These qualities, combined with a low melting point, have made it a popular material for 3D printing, alongside other manufacturing processes such as injection moulding [10].

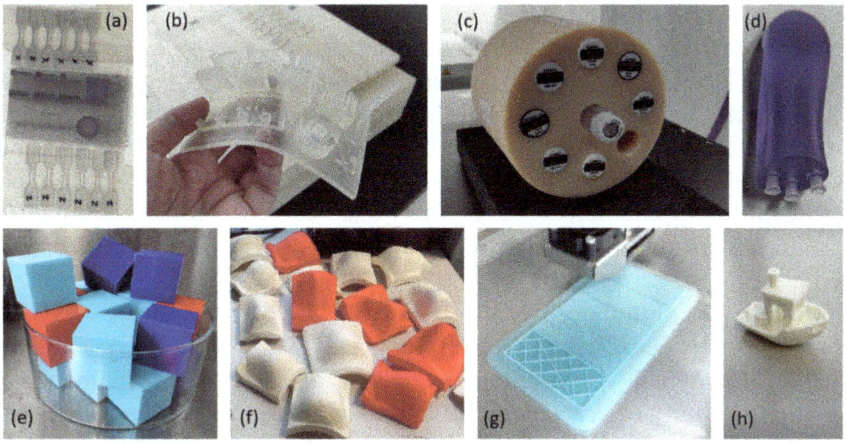

Figure 3.1. Examples of 3D-printed plastics for radiation therapy applications: (a) test samples of photopolymers; (b) test sample of flexible photopolymer; (c) large photopolymer print for use as a head-sized housing for commercial CT calibration plugs; (d) photopolymer model printed for evaluation as a brachytherapy applicator; (e), (f), (g), and (h) test samples of PLA printed with different densities and print parameters for respective evaluation as potential modular thoracic phantom blocks, radiation therapy bolus, a PET imaging phantom, and the standard '3Dbenchy' [7] geometric test object.

Table 3.1. Common polymers used in 3D printing for radiation therapy, categorised by type of material and 3D-printing system, and showing the section in chpater 2 where information on the printing method can be found.

Hardware	Material	Common materials	Section
MEX	Filament	PLA, TPU, HIPS, ABS, PMMA, PC, PETG, PEEK	2.3
MJP	Resin	Agilus, Vero, VisiJet, MED 610	2.7.1
SLA	Resin	Somos WaterShed, Dental SG, BioMed, KeyGuide	2.2.1
LOM	Sheet	Plastic, paper, composites	2.6
DLP	Resin	R5 Resin, Varseo Wax Splint	2.2.2
SLS	Powder	PA2200, PA11, PA12 (Polyamide powders)	2.4.2
BJT	Powder	EOS PA 3200 (Polyamide powder)	2.8

ABS = acrylonitrile butadiene styrene, HIPS = high-impact polystyrene, PC = polycarbonate, PEEK = polyether ether ketone, PETG = polyethylene terephthalate glycol, PLA = polylactic acid, PMMA = polymethyl methacrylate, TPU = thermoplastic polyurethane. The 3D printer initialisms are defined in chapter 2.

Some ABS plastics, such as ABS M30i and ABS Medical Smartfil, are also used for medical applications [11, 12], including fabrication of scaffolds for bone and tissue regeneration applications [11, 13, 14] (see chapter 11). Due to this range of properties, ABS has found several applications in radiation therapy, such as making phantoms [15, 16], bolus [1, 17], and immobilisation devices for patients [8]. The main disadvantage of using ABS for printing within a clinical environment is its increased production of volatile organic compounds when compared with alternative materials such as PLA (see chapter 6).

PLA is another commonly-used polymer for MEX printing [18]. PLA is an organic polymer that is derived from renewable resources such as potato, corn, and sugarcane, which has a significant potential to replace the petroleum-based polymers due to its strength, natural biocompatibility, and low toxicity [9]. It is reported in the literature that PLA is one of the most commonly-used synthetic polymers for medical applications [19], and its low cost and excellent properties make it effective in many medical applications [20]. PLA is also one of the most frequently-used plastic materials for manufacturing bolus and immobilisation devices for radiation therapy [1, 8].

Polyethylene terephthalate glycol (PETG) has also found use for 3D-printing radiation therapy devices. Polyethylene terephthalate glycol (PETG) is an industrial strength filament, which has combined the ease of use of PLA filament with the strength and durability of ABS filament. With a strength higher than PLA, PETG is suitable for a wide range of applications, including radiation therapy.

While the previous materials can be quite rigid, thermoplastic polyurethane (TPU) has flexible mechanical properties that can be advantageous for medical applications [21]. The TPU polymer chain comprises two segments: the soft segments, which consist of flexible polyethers or polyester; and hard segments, which consist of diisocyanates with benzyl construct [22]. These chemical properties result in a material that can be highly flexible when 3D printed with low infill densities, or as thin or narrow structures, and yet is substantially more rigid when used to print more substantial objects at higher densities [23].

The bio-stability, ease of fabrication of instruments, electrical insulation properties, flexibility, and biocompatibility of TPU has made it suitable for a variety of medical applications [24]. Additionally, TPU has good tear resistance and abrasion, moderate tensile strength, moderate Young's modulus, and high elongation [21]. However, TPU has some drawbacks, such as poor mechanical strength and shape fixity [22], and when printed using MEX extrusion technologies can require modified hardware and careful tuning of process parameters in order to work. Powder bed fusion (PBF) technologies can also be used for 3D printing with TPU (see chapter 2).

Polyether ether ketone (PEEK) is a linear semi-crystalline polycyclic aromatic thermoplastic and is a common thermoplastic material in radiation therapy, including in commercial products such as vaginal domes used in brachytherapy. PEEK has significant mechanical resistance properties, but is a costly thermoplastic material [18] and requires the use of high temperatures and heated enclosures for 3D printing. PEEK has good strength and stiffness and exhibits excellent biocompatibility [25], which make it desirable for many surgical and medical applications [26]. Like ABS, PLA, and TPU, PEEK is safe to image with MRI [27], is radiolucent, and is chemically stable. PEEK can also be made biocompatible [28].

A wide variety of thermoplastic 3D-print materials are nominally biocompatible but not all are suitable for sterilisation without material deformation. Biocompatible polycarbonate (PC-ISO), polyphenylsulphone (PPSF), and Ultem 9085 can be sterilised, and have been reported in use as 3D-printed custom brachytherapy applicators [29]. In the literature, 3D-printing materials such as PLA and ABS to

TPU, polymethyl methacrylate (PMMA), and high-impact polystyrene (HIPS) are used to manufacture anthropomorphic phantoms [1] (see chapter 10).

The MEX materials described so far, other than PEEK, are relatively inexpensive, and so are the most commonly used for radiation therapy devices. For objects that must be biocompatible and sterilisable, such as intracavitary brachytherapy applicators, there are many options available using other 3D-printing technologies, including photopolymer resins such as the Formlabs BioMed and KeyPrint KeyGuide product ranges for vat photopolymerisation (VPP) 3D printing or MED610 for material jetting (MJT) printing, or Nylon 12 polyamide for selective layer sintering (SLS) printing [30]. Objects made with photopolymer resins may be able to be heated or chemically sterilised, and have been extensively used within dentistry [31].

The advantages of plastics include resistance to corrosion, ease of processing, lightness, resilience, and transparency [2]. Plastics also come in a range of colours, and through processes such as multi-jet fusion (MJF) or multi-jet printing (MJP) allow for full-colour 3D printing (see chapter 2), which may provide opportunities to improve aesthetics or provide additional features such as markers, patient details, or other information. Due to the carbon backbone of polymers, the effective atomic number of most polymers is close to that of water. Additionally, most 3D-printable polymers have a similar density to water and tissue, which places their effective mass attenuation coefficient at approximately the same value as water and tissue.

Plastics also have several limitations, including mechanical strength and flexibility, the effect of heat (for thermoplastics) or ultra-violet exposure (for thermosets), and the limited range of radiological properties. Porosity is a challenge for 3D-printed plastic medical devices, which is caused by the layer-by-layer nature of the technology, as well as material characteristics and methods of fusion. This can affect performance and also provide areas for bacteria to grow. Chapter 6 describes the costs associated with plastics, including challenges of recycling and disposal [32] and the production of volatile organic compounds during the 3D-print process [33, 34].

The material properties of polymers are significantly affected by several factors, including the polymerisation technique of the monomer and extrusion (e.g. thermoplastic vs thermosetting polyethylene, PE), the functional groups on the carbon backbone (e.g. polyvinyl alcohol, PVA, vs polyvinyl chloride, PVC), and molecular weight (polymer chain length). For MEX printing, thermal degradation of a polymer will occur at the melting point because there is a transition from solid to liquid. Bonds in the polymer are broken when transitioning from the solid to liquid phase, which influence the material properties, such as reduced strength. Since the process of 3D printing extrudes polymers at temperatures close to their melting point, some thermal degradation is always present. There are three main instances in which thermal degradation can occur: by purely thermal means, thermal-oxidative, or thermal-mechanical [35]. Depending on the density and structure of the polymer, the elastic properties can be altered significantly. Thermosetting polymers cannot be remoulded once they are heated and then subsequently cooled. Hence, they are only used in stereolithography and direct writing 3D-printing techniques, which require a combination of heating and ultra-violet or visible photocuring [36] (see chapter 2). Thermoset polymer resins will harden and become more brittle compared to thermoplastic polymers, which remain quite plastic even after cooling and can also be reheated after setting. The elastic modulus of most 3D-printable polymers remains between 1 and 4 GPa but can be reduced by the effects of infill and layer adhesion [37].

Material degradation can occur from exposure to ultra-violet light, interaction with chemicals such as alcohol and acetone, deterioration from excessive heat and mechanical stress, and biodegradation. Plastics can also be degraded by exposure to ionising radiation, which has led several authors to specifically evaluate the effects of therapeutic doses of radiation on 3D-printed plastics intended for radiation therapy use [38–40].

For the most common thermoplastic (i.e. 3D-printable polymers such as ABS and PLA), the plastic will warp and deteriorate when interacting with alcohol or acetone. For ABS, alcohol and acetone will liquify the polymer but in small amounts can be used to smooth the print. While PLA is more resistant to these chemicals, Lye (NaOH) will 'eat away' at the polymer [36]. Hence, PLA-based support can be stripped using Lye. ABS, PLA, and PETG are unstable when exposed to acids or bases; however, for specifics, a chemical or material database should be consulted. Some 3D-printable polymers, such as Nylon and PEEK, exhibit greater chemical resistance to some inorganic compounds such as ammonia and sulphuric acid but can still be oxidised by a basic halogenated solution.

Some of the limitations affecting plastics can be addressed by combining polymers with inorganic material to form 3D-printable composite materials, as described in section 3.3.

3.3 3D-printed composites

Composite materials are formed from a combination of two or more materials, with the final product often possessing unique properties. Plastic 3D-printing materials can be modified through the addition of other materials to achieve a range of desired properties, in some cases addressing the limitations of plastic through the addition of carbon fibre, such as increasing the stiffness and strength of an object while maintaining light weight,. Some examples of polymer composite materials used in 3D printing for radiation therapy applications are shown in figure 3.2.

The radiological properties of plastics can be altered through the addition of other materials to achieve effective mass and electron densities matching those of different human tissues, which makes them suitable for imaging and therapy phantoms [15, 41]. For example, commercially-available PLA composites (including wood fibre or chemical foaming agents) print at lower densities than unmodified PLA [42–44] and can be combined with appropriate print parameters to achieve lung equivalency in 3D-printed phantoms. The radiological characteristics of denser tissues including bone can be modelled more or less accurately by using PLA composites containing bone, powdered stone. Metals (including aluminium, brass,

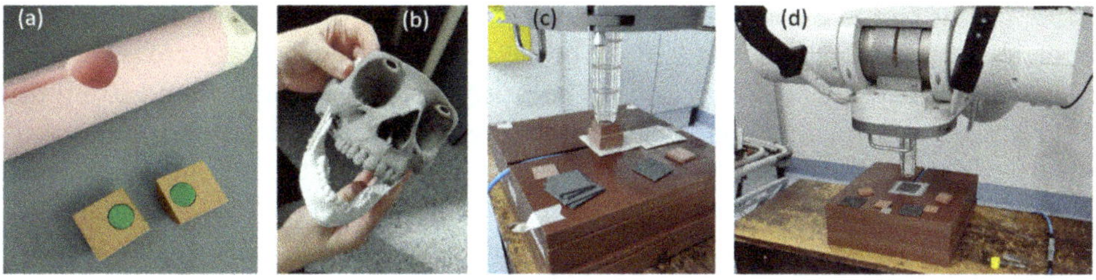

Figure 3.2. Examples of 3D-printed composites, for radiation therapy applications: (a) a wood-PLA composite (bottom of image) for modelling a lung as a custom insert for a commercial thoracic phantom (top of image); (b) a stone-PLA composite for modelling bone (bone-equivalent at megavoltage photon treatment energies) in a skull phantom with a mobile mandible; and (c) copper-PLA and (d) a tungsten-PLA composites, shown being tested for utility as potential kV radiation therapy shielding materials.

copper, iron, steel, and tungsten) can be used to mimic higher density tissues such as bone in phantoms [15, 42, 43, 45] (see chapter 10). PLA composites containing copper or tungsten can also be used to produce patient shielding for superficial radiation therapy treatments [46, 47] (see chapter 9). These techniques are not limited to MEX composites: the StrataSys RadioMatrix MJP material allows varying concentrations of a radio-pacifier to be used for contrast of x-ray imaging phantoms.

In addition to the commercially-available composite materials, composites can also be manufactured by the user. In the context of 3D printing, most thermoplastic polymers are extruded to long fibrous filaments to be utilised for MEX and SLS printing [36]. However plastics materials can be purchased in powder and pellet form, from which filaments can be produced. These filaments can be doped with additives such as radio-pacifiers (e.g. barium sulphate), to achieve the desired material properties [45]. Similarly, materials can be added to photopolymerising resins prior to printing, as demonstrated by Badiuk *et al* [48], who used powdered calcium carbonate to achieve bone-equivalence in a maxillofacial phantom and Gillett *et al* [49] who used ^{18}F-FDG to produce a positron emission tomography (PET)-visible phantom. Plastics can be also combined with conductive materials such as graphitised carbon or graphene.

When using any 3D-printed material, it is important to understand that the material properties of the printed object will usually differ from the material properties of the feedstock material. This discrepancy is especially important to consider when dealing with composite materials, where changes in dispersion may cause material properties to vary throughout the print.

3.4 3D-printed metals

Metals can be 3D printed using methods such as PBF and direct energy deposition (DED) [50, 51], as well as their subcategories (i.e. MJF, SLS, DMLS, SLM, EBM, LENS, and EBAM), as described in chapter 2.

The unique chemical and physical properties of metals differentiate them from polymers and other materials. Metals have high melting temperatures, are good thermal and electrical conductors, possess high densities, and are malleable and ductile (with some exceptions). In load bearing applications, the mechanical properties of metallic structures must be sufficient. When metals are 3D printed, this characteristic may be compromised because the elastic modulus and strength are often reduced as a result of the weaker bonding between particles and the increased porosity of the bulk structure [52]. In particular, the ductility and tensile strength of 3D-printed materials decrease compared to the bulk materials. However, this can be dependent on the printing parameters, such as temperature, because this will affect the fusion of the particles and spacing [52, 53]. Recent studies have indicated that it is possible to match the properties obtained from traditional methods, as demonstrated by Sun *et al* [54] where the properties of stainless steel were improved by customising the crystallographic texture.

The thermal properties of metals are an important quantity that affects the manufacturing process of the material. For selective layer melting (SLM) printing, the thermal conductivity of the powder will affect the temperature distribution, and can limit the size and microstructure of the components [55]. The actual thermal conductivities can be influenced significantly by the packing density of the powder particles in the print bed, with denser packing allowing greater contact between each individual particle and improving the heat transfer [55].

In choosing metals in terms of conductivity, silver and gold have some of the lowest resistivities but are too expensive for anything other than thin films. Copper is widely approved for electrical appliances but is difficult to 3D-print due to oxidation, high thermal conductivity, reflectivity, and unstable melting [56]. Recent advancements have made printing copper possible, either through indirect SLM by mixing with polymethyl methacylate as a feedstock or through direct SLM by using a CO_2 laser, limiting the remission from the reflective copper [56, 57].

While many metal alloys are able to be 3D printed [6], the 3D printing of elemental metals with high atomic numbers such as lead or tungsten is not currently feasible, except for small sizes and thin layers using a technique known as laser-induced forward transfer [6]. Through the use of plastic composites, as described in section 3.3, it possible to achieve qualities similar to metals (e.g. conductivity, radiation shielding), without the weight or cost of 3D printing directly with elemental metals. Additionally, metallic composites printed via MEX may further be improved in terms of mechanical properties via heat furnacing after printing to dissolve remaining polymeric and binding structures, and fuse the metallic structures, which makes the objects denser and pore-free.

3.5 Other materials

3D printing can be used to create moulds, positive models for conventional moulding, and objects containing compartments for other materials. These techniques are sometimes contrasted against the 'direct' 3D printing of an object as 'indirect' 3D printing or manufacturing [45, 58]. Indirect 3D printing allows the use of materials that cannot be directly 3D printed, combinations of materials that cannot be 3D printed simultaneously (e.g. by VPP, PBF or a MEX printer with a single nozzle), and materials that are more cost-effective or flexible than

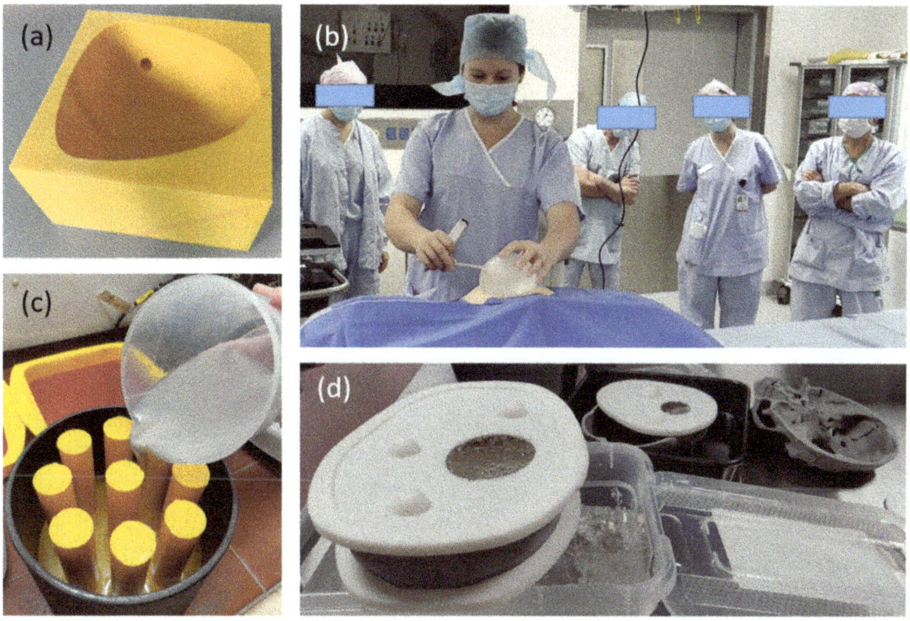

Figure 3.3. Example applications of 3D-printed moulds to create phantoms for radiation therapy applications: (a) plastic 3D-printed mould to create a gel phantom for intra-operative radiation therapy dosimetry and training; (b) physicist provides training to surgical staff using a breast phantom produced using (a); (c) 3D-printed mould for setting resin to act as head-sized CT calibration phantom; and (d) 3D-printed mould sections to create an MRI-compatible phantom for brain-metastases radiation therapy end-to-end testing.

3D-printable alternatives. Figure 3.3 shows some of the uses of simple 3D-printed plastic moulds (printed in PLA using MEX) to create phantoms from resin and various gels.

3D printing can be used to make negatives to cast patient-matched devices (described in chapter 9), potentially using materials with more desirable properties than printable alternatives, such as silicone for radiation therapy bolus [59, 60]. Electron field shaping devices have long been moulded using high density materials, usually from an alloy of lead and cadmium such as Cerrobend [61]. There is potential to 3D-print moulds for these materials to accommodate any field shape (see chapter 9). Skinner *et al* [61] demonstrated an alternative to this using a 3D plastic mould that was then filled with tungsten ball bearings, yielding similar beam profiles within 1.2% compared to casted electron beam cut-outs. Leat *et al* [62] described a similar solution using tungsten carbide.

Alternatively, 3D printing can be combined with 3D surface scanning and used to make positive models, over which lead shielding for kilovoltage photon radiation therapy can be formed, as an alternative to the uncomfortable and labour intensive practice of creating moulds directly on the patient's skin using plaster or thermoplastic [63, 64]. Work in this area is described in chapter 9.

MRI-visible phantoms contain compartments or voids that are later filled. While there are some 3D-printable materials that are MRI-visible [65], most 3D-printed

MRI-visible phantoms feature compartments that are filled with water, oils, and agarose or agar-based gels to achieve MRI visibility [58]. Similar approaches are taken with phantoms for single photon emission computed tomography or PET imaging by filling compartments with fluid radiotracers [58]. Gels, waxes, resins, silicones, and rubbers can be used in place of 3D-printed materials as tissue surrogates in radiation imaging or dosimetry phantoms, reducing filament use and 3D-printing times. For example, Badiuk *et al* [48] described the use of polyurethane rubber and epoxy resin as tissue surrogates to fill in spaces within a maxillofacial phantom. These materials can be combined with additives to decrease or increase (kilovoltage) radiological density, such as glass beads [48] and barium sulphate [66], respectively.

Care should be taken when selecting 3D-printer materials to be used for mould materials that are heated (e.g. some gels or alloys) or that cure via exothermic reaction (e.g. some resins), due to the potential of deformation of the printed mould. This is of particular concern when using common MEX printed thermoplastics that may be deformed by the heat released by the moulded material. The use of liquids in moulds may require surface post-processing using sealants or chemical smoothing to avoid layer separation due to pressure applied by the moulded material as it sets.

3.6 Key points

- Polymers (commonly referred to as plastics) are the most commonly-used materials for 3D printing in radiation therapy.
- For the most common form of 3D printing, MEX, water-equivalence can be achieved using inexpensive commercially-available plastic filaments, including ABS, PLA, and PETG, as well as the currently more costly PEEK.
- The production of biocompatible and sterilisable objects or devices may warrant the use of VPP and MJT printers, and the specialist materials available on those platforms.
- Metals can also be printed using PBF and DED systems (see chapter 2), although printing of high atomic number metals is only achievable in limited volumes, except by combining the metals with plastics to make composite materials.
- MEX (and some MJP) 3D printing can use composite materials to produce radiation therapy and imaging phantoms with water-, lung-, and (to varying degrees) bone-equivalent regions.
- User-made composites and indirect manufacturing of moulds can be used to overcome the limitations in the properties of 3D-printable materials.

References

[1] Tino R, Leary M, Yeo A, Kyriakou E, Kron T and Brandt M 2020 Additive manufacturing in radiation oncology: a review of clinical practice, emerging trends and research opportunities *Int. J. Extreme Manuf.* **2** 012003

[2] Crawford R J and Martin P 2020 *Plastics Engineering* (Portsmouth, NH: Heinemann)

[3] Gibson I 2015 *Additive Manufacturing Technologies 3D Printing, Rapid Prototyping, and Direct Digital Manufacturing* (New York: Springer)

[4] Bîrcă A, Gherasim O, Grumezescu V and Grumezescu A M 2019 Introduction in thermoplastic and thermosetting polymers *Materials for Biomedical Engineering* (Amsterdam: Elsevier) pp 1–28

[5] Loy J and Novak J I 2021 Additive manufacturing for a dematerialized economy *Sustainable Manufacturing and Design* (Amsterdam: Elsevier) pp 19–45

[6] Ni J *et al* 2019 Three-dimensional printing of metals for biomedical applications *Mater. Today Bio* **3** 100024

[7] Creative Tools 2015 *#3dbenchy—The Jolly 3D Printing Torture-Test by creativetools.se* [Online] Available: https://www.thingiverse.com/thing:763622 (visited on 05/30/2023)

[8] Asfia A, Novak J I, Mohammed M I, Rolfe B and Kron T 2020 A review of 3D printed patient specific immobilisation devices in radiotherapy *Phys. Imaging Radiat. Oncol.* **13** 30–5

[9] Bijarimi M, Shahadah N, Ramli A, Nurdin S, Alhadadi W, Muzakkar M Z and Jaafar J 2020 Poly (lactic acid)(PLA)/acrylonitrile butadiene styrene (ABS) with graphene nanoplatelet (GNP) nanocomposites *Indonesian J. Chem.* **20** 276–81

[10] Srinivasan R, Rathish R, Sivaraman P, Pramod A and Shivaganesh G 2020 Influential analysis of fused deposition modeling process parameters on the wear behaviour of ABS parts *Mater. Today: Proc.* **27** 1896–76

[11] Rosenzweig D H, Carelli E, Steffen T, Jarzem P and Haglund L 2015 3d-printed ABS and PLA scaffolds for cartilage and nucleus pulposus tissue regeneration *Int. J. Mol. Sci.* **16** 15118–35

[12] Moraru E, Besnea D, Spanu A and Constantin V 2018 Experimental research on some modern technologies and materials applied in prosthetic dentistry *J. Eng. Sci. Innov.* **3** 329–38

[13] Mozdzen L C, Rodgers R, Banks J M, Bailey R C and Harley B A 2016 Increasing the strength and bioactivity of collagen scaffolds using customizable arrays of 3D-printed polymer fibers *Acta Biomater.* **33** 25–33

[14] Alaboodi A S and Sivasankaran S 2018 Experimental design and investigation on the mechanical behavior of novel 3D printed biocompatibility polycarbonate scaffolds for medical applications *J. Manuf. Process* **35** 479–91

[15] Kairn T, Zahrani M, Cassim N, Livingstone A, Charles P and Crowe S 2020 Quasi-simultaneous 3D printing of muscle-, lung-and bone-equivalent media: a proof-of-concept study *Phys. Eng. Sci. Med.* **43** 701–10

[16] Tino R B, Yeo A U, Brandt M, Leary M and Kron T 2022 A customizable anthropomorphic phantom for dosimetric verification of 3D-printed lung, tissue, and bone density materials *Med. Phys.* **49** 52–69

[17] Sasaki D K, McGeachy P, Aviles J E A, McCurdy B, Koul R and Dubey A 2019 A modern mold room: Meshing 3D surface scanning, digital design, and 3D printing with bolus fabrication *J. Appl. Clin. Med. Phys.* **20** 78–85

[18] Arif M, Kumar S, Varadarajan K and Cantwell W 2018 Performance of biocompatible PEEK processed by fused deposition additive manufacturing *Mater. Des.* **146** 249–59

[19] Przybytek A, Gubańska I, Kucińska-Lipka J and Janik H 2018 Polyurethanes as a potential medical-grade filament for use in fused deposition modeling 3D printers—a brief review *Fibres Text. East. Eur.* **26** 120–5

[20] Hamad K, Kaseem M, Yang H, Deri F and Ko Y 2015 Properties and medical applications of polylactic acid: a review *Express Polym. Lett.* **9** 435–55

[21] Mi H-Y, Salick M R, Jing X, Jacques B R, Crone W C, Peng X-F and Turng L-S 2013 Characterization of thermoplastic polyurethane/polylactic acid (TPU/PLA) tissue engineering scaffolds fabricated by microcellular injection molding *Mater. Sci. Eng.* C **33** 4767–76
[22] Chen Q, Mangadlao J D, Wallat J, De Leon A, Pokorski J K and Advincula R C 2017 3D printing biocompatible polyurethane/poly (lactic acid)/graphene oxide nanocomposites: anisotropic properties *Appl. Mater. Interfaces* **9** 4015–23
[23] Zhao Y, Moran K, Yewondwossen M, Allan J, Clarke S, Rajaraman M, Wilke D, Joseph P and Robar J L 2017 Clinical applications of 3-dimensional printing in radiation therapy *Med. Dosim.* **42** 150–5
[24] Martin D, Osman A, Andriani Y and Edwards G 2012 Thermoplastic polyurethane (TPU)-based polymer nanocomposites *Advances in Polymer Nanocomposites* (Amsterdam: Elsevier) pp 321–50
[25] Vaezi M and Yang S 2015 Extrusion-based additive manufacturing of PEEK for biomedical applications *Virtual Phys. Prototyping* **10** 123–35
[26] Panayotov I V, Orti V, Cuisinier F and Yachouh J 2016 Polyetheretherketone (PEEK) for medical applications *J. Mater. Sci.: Mater. Med* **27** 1–11
[27] Ozdil D and Aydin H M 2014 Polymers for medical and tissue engineering applications *J. Chem. Technol. Biotechnol.* **89** 1793–810
[28] Ma R and Tang T 2014 Current strategies to improve the bioactivity of PEEK *Int. J. Mol. Sci.* **15** 5426–45
[29] Fowler T L, Buyyounouski M K, Jenkins C H and Fahimian B P 2016 Clinical implementation of 3D printing for brachytherapy: techniques and emerging applications *Brachytherapy* **15** S166
[30] Wilks R, Skewes J, Forrestal D, Kairn T and Crowe S 2023 Evaluating 3D printing materials for intracavitary gynaecological brachytherapy *Phys. Eng. Sci. Med.* **46** 482
[31] Andjela L, Abdurahmanovich V M, Vladimirovna S N, Mikhailovna G I, Yurievich D D and Alekseevna M Y 2022 A review on vat photopolymerization 3D-printing processes for dental application *Dental Mater.* **38** e284–96
[32] Peeters B, Kiratli N and Semeijn J 2019 A barrier analysis for distributed recycling of 3D printing waste: taking the maker movement perspective *J. Cleaner Prod.* **241** 118313
[33] Azimi P, Zhao D, Pouzet C, Crain N E and Stephens B 2016 Emissions of ultrafine particles and volatile organic compounds from commercially available desktop three-dimensional printers with multiple filaments *Environ. Sci. Technol.* **50** 1260–8
[34] Du Preez S, Johnson A, LeBouf R F, Linde S J, Stefaniak A B and Du Plessis J 2018 Exposures during industrial 3-D printing and post-processing tasks *Rapid Prototyping J.* **24** 865–71
[35] Li Y, Qiang Z, Chen X and Ren J 2019 Understanding thermal decomposition kinetics of flame-retardant thermoset polylactic acid *RSC Adv.* **9** 3128–39
[36] Sing S, Tey C, Tan J, Huang S and Yeong W Y 2019 Functional 3D printed polymeric materials *Functional Materials* (London: IntechOpen) pp 1–5
[37] Katsifis G A, McKenzie D R and Suchowerska N 2019 Applying the Hashin-Shtrikman bounds to predict stiffness of multicomponent 3D printed structures: towards regenerative orthopaedic medicine *J. Compos. Mater.* **54** 2173–83
[38] Kairn T, Crowe S and Markwell T 2015 Use of 3D printed materials as tissue-equivalent phantoms *World Congress on Medical Physics and Biomedical Engineering, June 7–12, 2015, Toronto, Canada* (Berlin: Springer) pp 728–31

[39] Van der Walt M, Crabtree T and Albantow C 2019 PLA as a suitable 3D printing thermoplastic for use in external beam radiotherapy *Australas. Phys. Eng. Sci. Med.* **42** 1165–76

[40] Michiels S *et al* 2016 Towards 3D printed multifunctional immobilization for proton therapy: initial materials characterization *Med. Phys.* **43** 5392–402

[41] McGarry C K *et al* 2020 Tissue mimicking materials for imaging and therapy phantoms: a review *Phys. Med. Biol.* **65** 23TR01

[42] Dancewicz O, Sylvander S, Markwell T, Crowe S and Trapp J 2017 Radiological properties of 3D printed materials in kilovoltage and megavoltage photon beams *Phys. Med.* **38** 111–8

[43] Savi M, Andrade M A B and Potiens M P A 2020 Commercial filament testing for use in 3D printed phantoms *Radiat. Phys. Chem.* **174** 108906

[44] Crowe S B, Maxwell S K, Brar H, Yu L and Kairn T 2023 Use of light-weight foaming polylactic acid as a lung-equivalent material in 3D printed phantoms *Phys. Eng. Sci. Med.* https://doi.org/10.1007/s13246-023-01318-4

[45] Tino R, Yeo A, Brandt M, Leary M and Kron T 2021 The interlace deposition method of bone equivalent material extrusion 3D printing for imaging in radiotherapy *Mater. Des.* **199** 109439

[46] Craft D F, Lentz J, Armstrong M, Foster M, Gagneur J, Harrington D, Schild S E and Fatyga M 2020 Three-dimensionally printed on-skin radiation shields using high-density filament *Pract. Radiat. Oncol.* **10** e543–50

[47] Crowe S B, Charles P H, Cassim N, Maxwell S K, Sylvander S R, Smith J G and Kairn T 2021 Predicting the required thickness of custom shielding materials in kilovoltage radiotherapy beams *Phys. Med.* **81** 94–101

[48] Badiuk S R, Sasaki D K and Rickey D W 2022 An anthropomorphic maxillofacial phantom using 3-dimensional printing, polyurethane rubber and epoxy resin for dental imaging and dosimetry *Dentomaxillofacial Radiol.* **51** 20200323

[49] Gillett D, Marsden D, Ballout S, Attili B, Bird N, Heard S, Gurnell M, Mendichovszky I A and Aloj L 2021 3D printing 18F radioactive phantoms for pet imaging *EJNMMI Phys.* **8** 38

[50] Duda T and Raghavan L V 2016 3D metal printing technology *IFAC-PapersOnLine* **49** 103–10

[51] Sing S, Tey C, Tan J, Huang S and Yeong W Y 2020 3D printing of metals in rapid prototyping of biomaterials: techniques in additive manufacturing *Rapid Prototyping of Biomaterials* (Amsterdam: Elsevier) pp 17–40

[52] Tan X, Kok Y, Tan Y J, Descoins M, Mangelinck D, Tor S B, Leong K F and Chua C K 2015 Graded microstructure and mechanical properties of additive manufactured Ti-6Al-4V via electron beam melting *Acta Mater.* **97** 1–16

[53] Frazier W E 2014 Metal additive manufacturing: a review *J. Mater. Eng. Perform.* **23** 1917–28

[54] Sun Z, Tan X, Tor S B and Chua C K 2018 Simultaneously enhanced strength and ductility for 3d-printed stainless steel 316L by selective laser melting *NPG Asia Mater.* **10** 127–36

[55] Alkahari M R, Furumoto T, Ueda T, Hosokawa A, Tanaka R and Abdul Aziz M S 2012 Thermal conductivity of metal powder and consolidated material fabricated via selective laser melting *Key Eng. Mater.* **523** 244–9

[56] Tran T Q *et al* 2019 3D printing of highly pure copper *Metals* **9** 756

[57] Lykov P A, Safonov E V and Akhmedianov A M 2016 Selective laser melting of copper *Mater. Sci. Forum* **843** 284–8

[58] Filippou V and Tsoumpas C 2018 Recent advances on the development of phantoms using 3D printing for imaging with CT, MRI, PET, SPECT, and ultrasound *Med. Phys.* **45** e740–60
[59] Canters R A, Lips I M, Wendling M, Kusters M, van Zeeland M, Gerritsen R M, Poortmans P and Verhoef C G 2016 Clinical implementation of 3D printing in the construction of patient specific bolus for electron beam radiotherapy for non-melanoma skin cancer *Radiother. Oncol.* **121** 148–53
[60] Chiu T, Tan J, Brenner M, Gu X, Yang M, Westover K, Strom T, Sher D, Jiang S and Zhao B 2018 Three-dimensional printer-aided casting of soft, custom silicone boluses (SCSBs) for head and neck radiation therapy *Pract. Radiat. Oncol.* **8** e167–74
[61] Skinner L, Fahimian B P and Yu A S 2019 Tungsten filled 3D printed field shaping devices for electron beam radiation therapy *PLOS One* **14** e0217757
[62] Leat C J, Ramachandran P and Langton C M 2020 Tungsten carbide and LMPA electron cutouts: comparison and validation using monte carlo modelling and measurement of dose *Biomed. Phys. Eng. Express* **7** 015007
[63] Sharma A *et al* 2018 Low-cost optical scanner and 3-dimensional printing technology to create lead shielding for radiation therapy of facial skin cancer: first clinical case series *Adv. Radiat. Oncol.* **3** 288–96
[64] Briggs M, Clements H, Wynne N, Rennie A and Kellett D 2016 3D printed facial laser scans for the production of localised radiotherapy treatment masks–a case study *J. Vis. Commun. Med.* **39** 99–104
[65] Rai R, Wang Y F, Manton D, Dong B, Deshpande S and Liney G P 2019 Development of multi-purpose 3D printed phantoms for MRI *Phys. Med. Biol.* **64** 075010
[66] How S, Banjade D, Crowe S, Dillon G and Skimmings A 2023 An anthropomorphic 3D printed inhomogeneity thorax phantom slab for SBRT commissioning and quality assurance *Phys. Eng. Sci. Med.* **46** 575–83

Chapter 4

Design

Rance Tino, Martin Leary, Gorgio Andrew Katsifis, James I Novak and Scott B Crowe

4.1 Introduction

The first step towards three-dimensional (3D) printing a specific object or device is establishing the design requirements and creating one or more 3D models that satisfy those requirements. These 3D models will then be prepared for printing, using the processes described in chapter 5. This chapter provides an overview of the approaches used by clinicians, engineers, and researchers for the preparation of 3D models of radiation therapy devices. For the purposes of this overview, three principal types of devices are considered: patient-matched medical devices (detailed in chapter 9), anatomical models or anthropomorphic phantoms (detailed in chapter 10), and other devices designed from schematics or measurements (also detailed in chapter 10).

These three types of objects differ in design approaches. For example, anthropomorphic phantoms often feature organic shapes with curved surfaces that are not easily parameterised, and are therefore often designed using anatomical images (e.g. see the workflow in figure 4.1). Conversely, devices such as quality assurance jigs often feature inorganic shapes with flat surfaces or defined curvatures, and are therefore designed using parametric design or hard surface modelling tools. Patient-matched devices often feature some combination of these two approaches. For example, a superficial brachytherapy mould may start with an organic model acquired from medical images. The mould is then designed using defined expansions, Boolean operations, and parametric modelling of catheter channels.

This chapter discusses the main methods of imaging that have been used in radiation therapy centres (medical or radiographic imaging, typically CT, and 3D optical surface scanning), the segmentation of imaging data, and the software used to modify or create and modify 3D models. This chapter also describes methods for creating and modifying models using design software.

doi:10.1088/978-0-7503-3907-0ch4

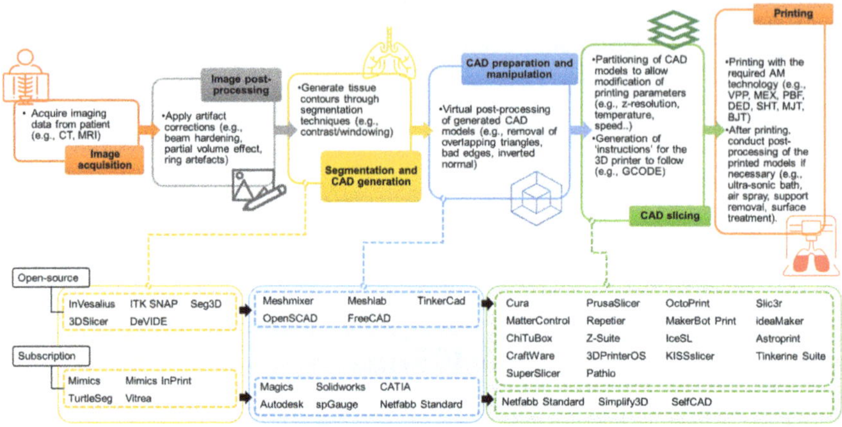

Figure 4.1. Common workflow for 3D printing of anatomical models and existing open-source and subscription-based software for segmentation, and computer-aided design preparation, manipulation, and slicing.

4.2 Medical imaging

Medical images of patients undergoing radiation treatments are acquired routinely and many radiation therapy centres have a dedicated CT scanner for the acquisition of treatment planning images. Many 3D-printed objects are designed to fit, accommodate, or replicate a specific device or anatomical site, which can be achieved through the use of treatment planning CT images (and images from other modalities, such as cone-beam CT images acquired for image guided radiation therapy, or diagnostic and planning MRI data, where available). CT images can be used as a starting point for the 3D-print design process, in addition to a print quality assurance verification tool, as described in chapter 8. 3D models can be created from images by segmentation, which is also known as contour delineation.

As with diagnostic imaging, tumour delineation, and treatment planning, the quality of the images used can limit the quality of the ultimate 3D-printing output (this is also an issue in other medical fields, as discussed in chapter 11). The ideal images will have a high resolution (small slice thickness and reduced field of view), high image quality (including an appropriate level of contrast between tissues and surrounding media for intended segmentation), and have minimal artefacts. When high precision scans of objects are required, orientation should be carefully considered because pixel spacing within a transverse slice will typically be smaller than the slice thickness. Micro-CT imaging systems may provide improved resolution if available. Where the density of the object being scanned is similar to that of the top of the CT couch, separating the two using a lower density material (e.g. a foam patient support or balsa wood) can simplify segmentation [1]. These CT scanning arrangements are exemplified in figure 4.2.

When scanning patients or objects containing high density materials such as metals, the use of metal artefact reduction algorithms or higher energy systems (e.g. megavoltage imaging on tomotherapy and Halcyon platforms) may improve image quality.

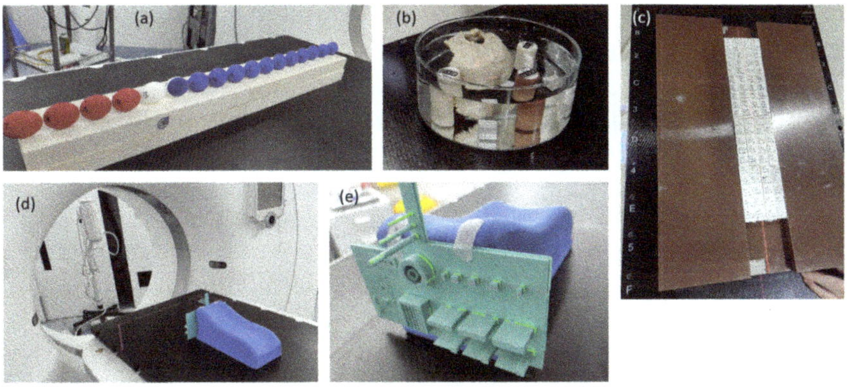

Figure 4.2. Examples of CT scanning arrangements for 3D-printed objects used in radiation therapy investigations: (a) samples placed on low density material (balsa wood) to achieve separation from CT couch; (b) samples immersed in water; c) samples surrounded with water-equivalent plastic to achieve clinically realistic CT beam hardening; and (d) and (e) test object positioned parallel to scan plane to maximise resolution in dimensions of interest.

The relationship between CT numbers (Hounsfield units, HU) and densities is routinely characterised for planning CT systems to allow assignment of mass density, electron density, and/or material composition for dose calculations. This relationship varies with degree of beam hardening, which itself varies with body size or water-equivalent thickness [2]. When scanning small objects, more accurate results can be obtained by surrounding the object with water or equivalent material. Some examples of useful CT scanning setups are shown in figure 4.2.

The output of medical imaging systems is produced in the form of voxelised 3D data sets, which are typically stored and communicated in compliance with the Digital Imaging and Communications in Medicine (DICOM) standard. While there are solutions that are able to directly print this voxelised data for phantom applications [3, 4], in most cases 3D models need to be prepared. Software enables the users to view medical imaging data and to segment different tissues through thresholding or other contouring techniques to generate 3D models for 3D printing.

Segmentation, or contour delineation, is well established with radiation therapy, where it is used to define target and organ-at-risk volumes, avoidance structures, and regions of density overrides and dose scoring for treatment planning. For these treatment planning workflows, this segmentation work is commonly done on either treatment planning systems (e.g. RayStation, Eclipse, and Monaco) or dedicated imaging contouring and registration software (e.g. MIM or Velocity). In addition to these commercial radiation oncology solutions, segmentation can be achieved with other commercial solutions (e.g. Materialise Mimics), open-source alternatives (e.g. 3D Slicer), or programmatically through the development of software in environments such as Python or MATLAB.

Several segmentation tools have been implemented within these solutions to assist with the delineation of 3D volumes of interest, including:

- Thresholding, which is the selection of all voxels within a specified range of CT numbers. CT numbers (or other pixel values) can also be used for seeding and flood filling techniques.
- Manual delineation within one or more slices (e.g. by voxel-by-voxel painting or lasso selection). Interpolation can be used for contouring between segmented slices.
- Margin expansion or contraction, where the segment can be grown or shrunk by a specified margin.
- Logical (Boolean) operators, such as addition, subtraction, and intersection.
- Island separation, which is the splitting of non-contiguous segments into multiple contiguous segments.
- Automatic segmentation using atlas-based and deep-learning models.

These tools enable specific structures within a medical image to be converted to 3D volumes. Figure 4.3 shows the segmentation of bone and lung tissue within the Varian Eclipse treatment planning system. For segmentation of interfaces between materials (including the object and surrounding air), the threshold should be centred between the two pixel values to minimise errors associated with volume averaging. For example, for a water-equivalent object (0 HU) scanned in air (−1000 HU), a threshold of −500 HU would be appropriate.

Depending on the software used for segmentation, the 3D volume may be stored as a voxelised mask or label map (indicating which voxels are inside the segmented volume) or a series of planar contours. Prior to 3D printing, this needs to be converted to a mesh that describes the external surface of the 3D model. This process commonly uses the marching-cubes algorithm [5], which stitches the generated 2D

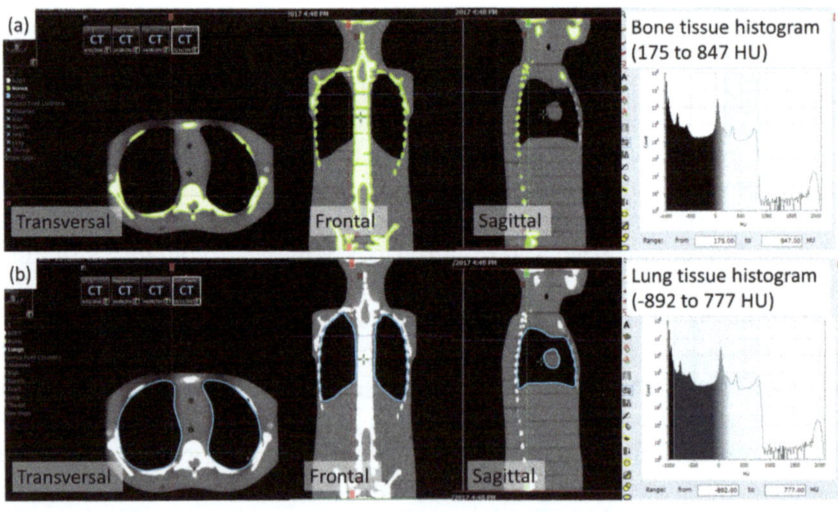

Figure 4.3. Segmentation of bone and lung tissue structures using the Eclipse treatment planning system. (a) generating bone tissue structure using an HU range of 175 to 847. (b) generating lung tissue structure using an HU range of −892 to 777.

tissue contours via the triangulation of vertices and edges to form an approximated 3D surface reconstruction of the tissue. The generated 3D surface reconstruction becomes the basis of the most widely used file format in the 3D-printing world, called STL (originally representing 'stereolithography', but now known as 'standard tessellation language' or 'standard triangle language'). This conversion is supported in many of the software packages mentioned earlier. Figure 4.4 illustrates the use of a script within the Eclipse treatment planning software to achieve this.

The main advantage of the use of medical imaging modalities is the ubiquity of these solutions, and their associated software and archiving systems, within radiation therapy centres. Treatment devices can be directly designed within treatment planning systems and assessed after printing using accessible CT scanning systems [6]. The design of patient-matched devices prior to CT simulation may be

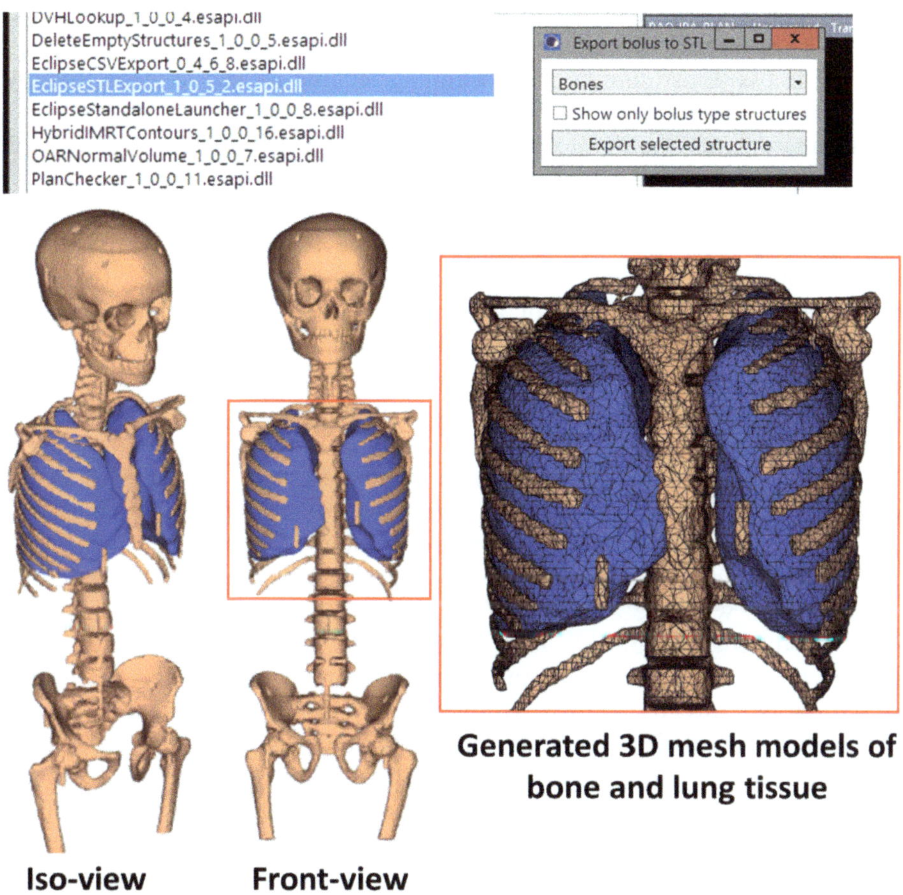

Figure 4.4. Converting generated tissue structure to a 3D mesh volume (.STL file format) using an in-house Python script integrated in Eclipse software. Generated mesh volumes were viewed through Materialise Magics software.

possible using diagnostic images or 3D optical surface scanning, which will be described in the next section.

It is important to note that the use of patient-specific data requires consent. In the case of research use or production of quality assurance tools, this is not necessarily established during the standard treatment consenting process. Consideration should also be given to secure storage, de-identification, and disposal of any STL data produced from patient medical images.

4.3 3D optical scanning

3D optical scanning involves the acquisition of topographical and appearance data to reconstruct a 3D model of a subject in the form of a point cloud or mesh, which can then be used for designing 3D-printed objects.

Optical surface imaging systems are widely used within a radiation therapy context for image guidance [7], and have been used in treatment planning and the production various patient-matched devices, including bolus, moulds, and shielding [6, 8–22]. The two most common 3D scanning techniques in radiation therapy are photogrammetry and structured light scanning [21], although alternatives such as laser scanning or lidar (i.e. light detection and ranging, found on some smartphone devices) may see increased adoption.

Photogrammetry involves the reconstruction of 3D models from multiple 2D photographs by building a cloud of points based on the identification and registration of common points (called landmarks) across multiple images, and the triangulation of the coordinates of these points from known or derived camera positions. These photographs may be acquired from a single fixed position, with the subject rotated to allow imaging of different aspects; from multiple fixed positions (e.g. with multiple cameras rigged in a surrounding frame); or from a moving camera, by an operator moving around a stationary subject. The photographs can be acquired on any system, from simple point-and-shoot cameras to expensive interchangeable-lens cameras, and to cameras integrated within mobile phones. The quality of the reconstruction will depend on number of photographs used and the number of common landmarks that can be identified in those photographs, which can be increased with the addition of markers on the subject (e.g. using lines or stickers). Producing an appropriately scaled model may require the addition of a calibration object to the scan, if camera positions relative to the subject (or other cameras) are not already known. Bridger *et al* [10] provided advice on camera settings to improve quality of models, including the use of low ISO (light sensitivity) values, low shutter speeds, and small aperture sizes. An example of a photogrammetric reconstruction using multiple cameras in a rig is shown in figure 4.5.

Structured light scanning involves the reconstruction of 3D models from multiple 2D images, in each of which a pattern of light has been projected onto the subject. The deformation of this pattern by the surface topography is then recorded. Similar to photogrammetry, more than one camera may be used, and either the subject or camera and projector system can be moved to acquire information from multiple aspects. Sophisticated structured light scanners are designed to be portable and are

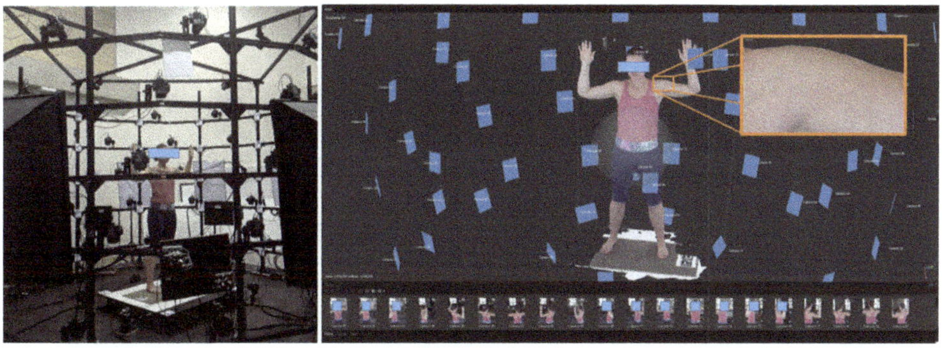

Figure 4.5. Example of photogrammetric 3D scanning of participant posing in a Stanford total skin electron therapy position, using a custom-made rig containing 76 interchangeable-lens cameras, with reconstruction using the Agisoft Metashape software. Inset shows achievable image quality.

metrology graded, providing detailed sub-millimetre precision models. These systems also frequently provide user feedback regarding scan progress (i.e. with coarse real time reconstructions that highlight what has and has not been imaged). These systems are, however, more complicated and expensive than photogrammetry solutions. An example of a structured light scanning using an Artec Leo device is shown in figure 4.6. This image was acquired as part of a study by Crowe *et al* [13], which included advice regarding the use of that scanner in a radiation therapy setting.

3D optical scanners have the potential to provide increased precision and accuracy compared to CT imaging, due to both the available resolution and the avoidance of threshold segmentation and volume averaging [11, 15, 22], which can improve the fit of patient-matched devices. Many 3D scanners capture surface textures in full colour, which allows the design of devices based on surface mark-ups of the visible target volume. The optical nature of these scanners has the potential to reduce dose associated with radiographic imaging (e.g. where devices need to be designed using patient imaging data) but also need be in-place at time of pre-treatment CT simulation. The geometric accuracy of the 3D scanning solution should be routinely verified, such as by imaging of objects of known dimensions [13].

When acquiring images of a human subject with a portable camera or scanner for the design of a treatment device, it may be convenient to have them sit in a comfortable position for ease of capturing different aspects. However, it is advisable to image them in the intended treatment position because surface anatomy can change due to the gravitational effects of moving from a seated (vertical) to a lying (horizontal) position [13]. This is particularly important if 3D optical scans are to be used for fixation or immobilisation devices. 3D scanning techniques are subject to motion artefacts, although this may be managed by partial reconstruction of motion binned data sets (e.g. before and after motion), and subsequent registration and stitching of the 3D models.

The data used for 3D scans can generally be acquired within minutes, but reconstruction can take longer. Imaging and reconstruction times scale with the size

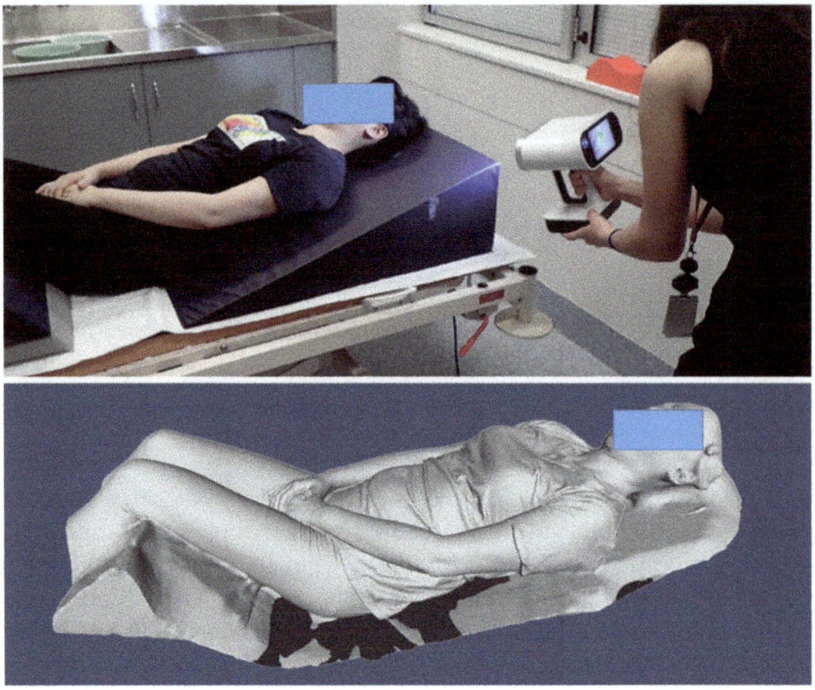

Figure 4.6. Example of structured light 3D scanning of participant using an Artec Leo device.

of the subject and area or interest. The occlusion or loss of line-of-sight of light patterns by a structured light scanner when imaging cavities or concave topographies may require the user to experiment with different scanner rotations. Reflective, shiny, wet, dark, and transparent surfaces, and very fine features such as hair can be difficult to image, depending on the technique used. Strategies for managing this include coating the surfaces with matte finish (e.g. with a powder or self-volatilising scanning spray).

The processing of photogrammetric or structured light scanning data may require substantial computational resources, such as a computer with a dedicated graphics processing unit (GPU). Some solutions, particularly those implemented on mobile phones, use cloud computing for the reconstruction of the 3D model, which may be inappropriate when handling patient data. The data sets associated with 3D scanning, including raw photographs or structured light images, and reconstructed models and textures, which can exceed gigabytes in size. Therefore, storage and archiving should be considered when planning the introduction of 3D optical scanning to the clinical workflow. The software used for reconstruction of photogrammetry or structured light scan data (including AliceVision Meshroom, Agisoft Metashape, 3DF Zephyr and Artec Studio) will handle the conversion of 3D point cloud data to meshes for 3D printing or importing into computer-aided design software. These meshes can also be converted to synthetic medical image data [12, 13, 15], which may be useful where design processes have been established within software (e.g. treatment planning systems) that do not necessarily support the import of point cloud or mesh data produced by 3D scanners. The use of 3D optical surface scanners for the design of radiation therapy treatment devices is discussed further in chapter 9.

4.4 Design software

The 3D models produced from the segmentation of CT images or 3D scanning may be immediately suitable for use in the production of phantoms, patient positives for moulding, or the replication of scanned devices. For other objects, these 3D models may need to be modified or entirely new 3D models may need to be created. There is a diverse range of design software tools to enable the design of objects or devices that meet user requirements. These tools can be categorised in a few ways, based on the type of design being done, whether direct soft (organic) or hard surface modelling, or parametric computer-aided design (CAD), and with varying degrees of ease-of-use and functionality needed to address design requirements.

To design simple objects or to make minor adjustments to existing models, hobbyist software such as Autodesk TinkerCAD, Autodesk Meshmixer, and Microsoft 3D Builder provide easy-to-learn hard surface modelling functionality, where logical operators can be combined with primitive volumes (cubes, cylinders, spheres) to create simple designs. Existing 3D models can be cut into parts and simple joins or keys can be added, with tolerances handled manually. Meshmixer and 3D Builder both provide similarly easy-to-use touch-up functionality, allowing smoothing and simplification by remeshing. The open-source software Blender provides similar functionality and additional features, but with a steeper learning curve, although this is mitigated by the availability of many online tutorials. Both Meshmixer and Blender also offer soft (or organic) surface modelling functions, such as pinching and smoothing. For 3D models produced from medical imaging data, repairs may sometimes need to be made (e.g. by filling voids, removing islands, or smoothing) prior to 3D printing. An example is shown in figures 4.7 and 4.8, where a skull has been segmented in 3D Slicer from a medical CT and converted into an STL file, displayed in Autodesk Meshmixer.

Segmentation is not usually perfect and will require further processing. In figure 4.8(a), the exported STL file of the skull possesses external features that are not connected to the skull. The segmented volume also has holes or voids, as seen predominantly in the eye sockets and around the nasal cavity/maxilla. To remove large islands, it is convenient to select each island and remove it. However for many small islands, selecting the large body and then inverting the selection to simply delete everything else is often more convenient. This is demonstrated in the workflow in figures 4.8(a)–(c). Figure 4.8(c) shows the two segmented volumes, one being the bulk skull structure and the other showing small pieces of disconnected tissue or bone. When there are many islands in the exported file, it may be necessary to readjust the thresholding techniques in the medical image processing software before exporting.

For small imperfections such as holes or cracks in the exported STL, these can be fixed with bridging, joining, or flood filled operations. In the workflow in figures 4.8(d)–(f), a set of cracks are selected on the nasal bone (figure 4.8(d)) and

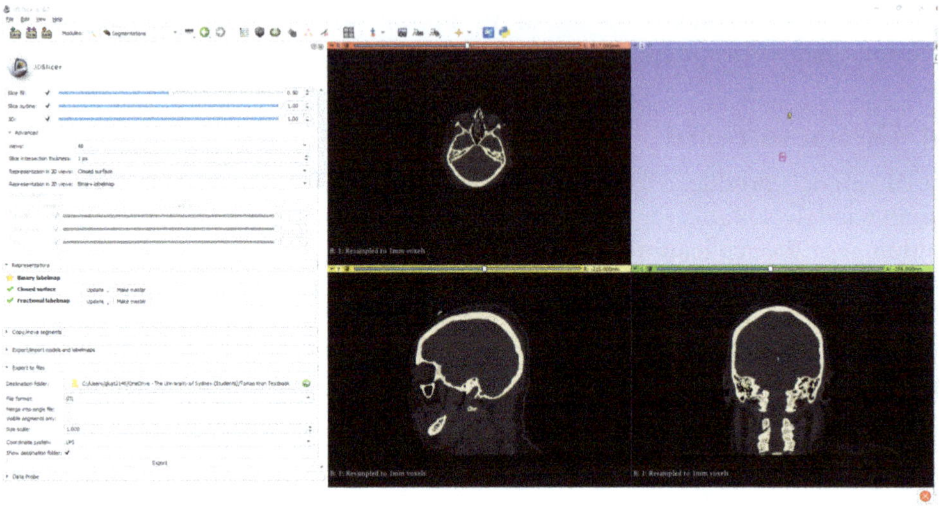

Figure 4.7. Skull bones in a cranial CT image are segmented using the segmentation module in 3D slicer.

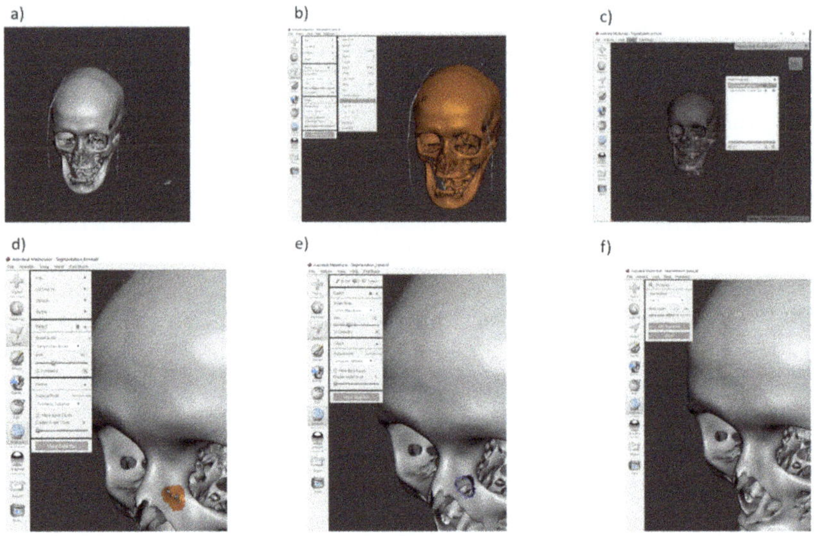

Figure 4.8. The segmented bone from 3D Slicer (figure 4.7) is exported to Autodesk Meshmixer in (a). In figures (b) and (c), the skull is selected and separated from the disconnected islands. Figures (d), (e), and (f) show the process of highlighting a hole (d), selecting and creating the defect (e), and then filling the defect (f).

are removed (figure 4.8(e)), leaving a hole in the region. This can then be repaired using the inspect function in Meshmixer, choosing the threshold and repairing the defect (figure 4.8(f)). For many holes, it is likely that different threshold values should have been used before exporting the segmentation as an STL.

To evaluate changes implemented from the original to the final model design, Meshlab can be utilised to assess geometric differences between the original and final

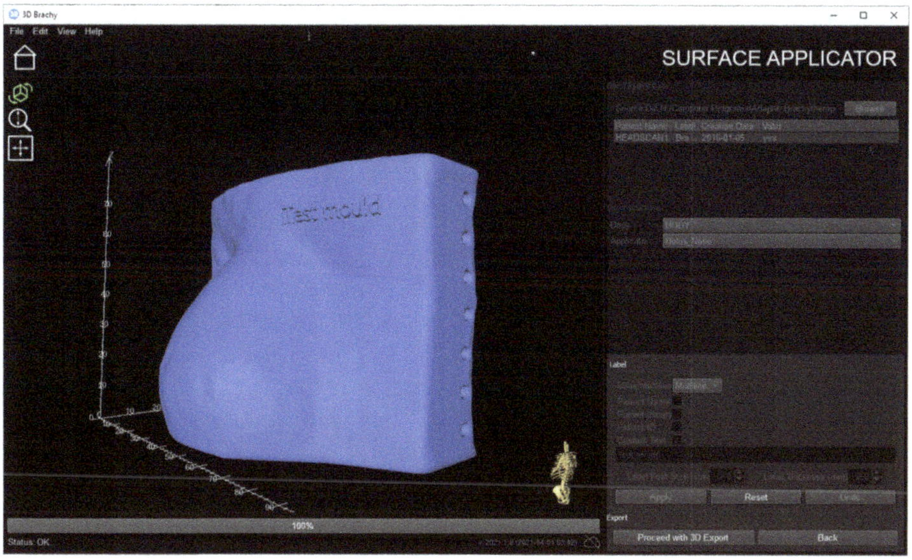

Figure 4.9. Screenshot of an applicator designed to treat the skin of the nose, from the 3D Brachy module of the Adaptiiv radiation therapy software suite, which handles the addition of catheter channels into brachytherapy moulds.

models via point-based registration method or the Hausdorff distance method [6] (see chapter 8, figure 8.5).

More advanced or parametric CAD tools may be required when designing more complex objects, such as Autodesk Fusion 360, SolidWorks, Rhino 3D, or open-source alternatives such as FreeCAD or OpenSCAD. These solutions require more time to learn but offer versatility, particularly when handling fitted parts, accounting for printer tolerances, and smoothing or chamfering edges.

For patient-matched devices, clinical treatment planning and imaging systems are often able to be used to design devices such as boluses or immobilisation equipment (e.g. through the use of margin expansions from the external or body contour). The virtual definition of these devices within the treatment plan can allow the dosimetric effect to be modelled and the geometric design to be optimised prior to 3D printing. Similar expansions are possible within software such as Autodesk Meshmixer by selection of a surface area (by surface painting or lasso selection), separation from the model, and extrusion away from the surface [13]. There may also be device-specific software solutions, such as Materialise Mimics for anatomical models, or the Adaptiiv suite for radiation therapy bolus and brachytherapy applicators (see figure 4.9).

4.5 Key points

- Medical images acquired for treatment planning are the most common source of 3D models for patient-matched devices and anthropomorphic phantoms.
- A wide range of software is available for 3D modelling for beginners to professionals, and many are available without cost.

- Segmentation (contour delineation) of regions (e.g. organs) in a CT data set can be performed using radiation therapy treatment planning software or other software using thresholding, Boolean operations, or automatic segmentation, as well as manual contouring methods.
- CAD tools can also be used to design simple geometric objects for 3D printing.
- For 3D printing purposes, segmented or otherwise designed structures usually need to be exported in STL format.
- STL files produced by segmenting medical images or 3D optical surface scans often need to be repaired to remove holes and delete islands prior to 3D print preparation.
- Care should be taken when producing, transferring, or archiving patient-specific 3D models to ensure that they are handled securely and are appropriately anonymised where possible.

References

[1] Kairn T *et al* 2021 Determining tolerance levels for quality assurance of 3D printed bolus for modulated arc radiotherapy of the nose *Phys. Eng. Sci. Med.* **44** 1187–99

[2] Zheng X, Al-Hayek Y, Cummins C, Li X, Nardi L, Albari K, Evans J, Roworth E and Seaton T 2020 Body size and tube voltage dependent corrections for Hounsfield Unit in medica X-ray computed tomography: theory and experiments *Sci. Rep.* **10** 15696

[3] Okkalidis N 2018 A novel 3D printing method for accurate anatomy replication in patient-specific phantoms *Med. Phys.* **45** 4600–6

[4] Okkalidis N and Bliznakova K 2022 A voxel-by-voxel method for mixing two filaments during a 3D printing process for soft-tissue replication in an anthropomorphic breast phantom *Phys. Med. Biol.* **67** 245019

[5] Fang L, Liu J, Liu J and Mao R 2018 Automatic segmentation and 3D reconstruction of spine based on FCN and marching cubes in CT volumes *2018 10th Int. Conf. on Modelling, Identification and Control (ICMIC)* pp 1–5

[6] Sasaki D K, McGeachy P, Aviles J E A, McCurdy B, Koul R and Dubey A 2019 A modern mold room: Meshing 3D surface scanning, digital design, and 3D printing with bolus fabrication *J. Appl. Clin. Med. Phys.* **20** 78–85

[7] Al-Hallaq H A *et al* 2022 AAPM Task Group report 302: surface-guided radiotherapy *Med. Phys.* **49** e82–112

[8] Jenkins C, Xing L and Yu A 2017 Using a handheld stereo depth camera to overcome limited field-of-view in simulation imaging for radiation therapy treatment planning *Med. Phys.* **44** 1857–64

[9] Yoshida E J *et al* 2020 Capturing 3D patient features for rapid prototyping in radiotherapy prior to simulation *J. Radiat. Oncol.* **9** 75–80

[10] Bridger C A, Douglass M J, Reich P D and Santos A M C 2021 Evaluation of camera settings for photogrammetric reconstruction of humanoid phantoms for EBRT bolus and HDR surface brachytherapy applications *Phys. Eng. Sci. Med.* **44** 457–71

[11] Bellis R, Rembielak A, Barnes E A, Paudel M and Ravi A 2021 Additive manufacturing (3d printing) in superficial brachytherapy *J. Contemp. Brachyther* **13** 468–82

[12] Skinner L, Knopp R, Wang Y-C, Dubrowski P, Bush K K, Limmer A, Trakul N, Million L, Marquez C M and Yu A S 2021 CT-less electron radiotherapy simulation and planning with a consumer 3D camera *J. Appl. Clin. Med. Phys.* **22** 128–36

[13] Crowe S *et al* 2022 Evaluation of optical 3D scanning system for radiotherapy use *J. Med. Radiat. Sci.* **69** 218–26

[14] Douglass M J J 2022 Can optical scanning technologies replace CT for 3D printed medical devices in radiation oncology? *J. Med. Radiat. Sci.* **69** 139–42

[15] Douglass M, Gorayski P, Patel S and Santos A 2023 Synthetic cranial MRI from 3D optical surface scans using deep learning for radiation therapy treatment planning *Phys. Eng. Sci. Med.* **46** 367–75

[16] Park J W, Oh S A, Yea J W and Kang M K 2017 Fabrication of malleable three-dimensional-printed customized bolus using three-dimensional scanner *PLOS One* **12** e0177562

[17] Dipasquale G, Poirier A, Sprunger Y, Uiterwijk J W E and Miralbell R 2018 Improving 3D-printing of megavoltage x-rays radiotherapy bolus with surface-scanner *Radiat. Oncol.* **13** 1–8

[18] Sharma A *et al* 2018 Low-cost optical scanner and 3-dimensional printing technology to create lead shielding for radiation therapy of facial skin cancer: first clinical case series *Adv. Radiat. Oncol.* **3** 288–96

[19] Douglass M J and Santos A M C 2019 Application of optical photogrammetry in radiation oncology: HDR surface mold brachytherapy *Brachytherapy* **18** 689–700

[20] LeCompte M C, Chung S A, McKee M M, Marshall T G, Frizzell B, Parker M, Blackstock A W and Farris M K 2019 Simple and rapid creation of customized 3-dimensional printed bolus using iPhone X true depth camera *Pract. Radiat. Oncol.* **9** e417–21

[21] Kang D, Wang B, Peng Y, Liu X and Deng X 2020 Low-cost iPhone-assisted processing to obtain radiotherapy bolus using optical surface reconstruction and 3D-printing *Sci. Rep.* **10** 8016

[22] Maxwell S, Charles P, Cassim N, Kairn T and Crowe S 2020 Assessing the fit of 3D printed bolus from CT, optical scanner and photogrammetry methods *Phys. Eng. Sci. Med.* **43** 601–7

Chapter 5

Processing

Rance Tino, Amirhossein Asfia, Giorgio Andrew Katsifis, James I Novak and Scott B Crowe

5.1 Introduction

The realisation of a virtual model (designed according to processes detailed in chapter 4) into a 3D-printed radiation therapy device in a safe, efficient, and reproducible way requires preparation that is specific to the hardware (chapter 2) and materials (chapter 3) being used. This chapter discusses the print preparation process, in addition to the process of starting and successfully completing a 3D print, as well as the post-processing steps that are often required to improve the appearance, function, and usability of the 3D-printed object.

The figures in this chapter are designed to illustrate the effects that 3D-print parameter selection can have on material extrusion (MEX, also known as FDM or FFF) printing results in particular, because MEX is the most common form of printing at point-of-care in radiation therapy centres [1–3] (see chapter 2). However, this text is mostly generalisable to other 3D-printing systems and materials, although notable differences for vat photopolymerisation (VPP), multi-jet printing, and sheet lamination (SHL) printing are acknowledged when they occur.

5.2 Preparation

5.2.1 Slicing

Prior to 3D printing, generated 3D models undergo a process called 'slicing', where a 3D model is converted into instructions for the 3D printer to follow. The most common format for printer instructions is known as 'g-code', which has been used for computer numerical control in computer-aided design and manufacturing (including subtractive manufacturing processes such as milling) since the 1960s. For 3D printing, these instructions contain Cartesian coordinates that specify where material should be extruded or deposited (i.e. x, y, and z), as well as the amount of material, layer-by-layer.

doi:10.1088/978-0-7503-3907-0ch5

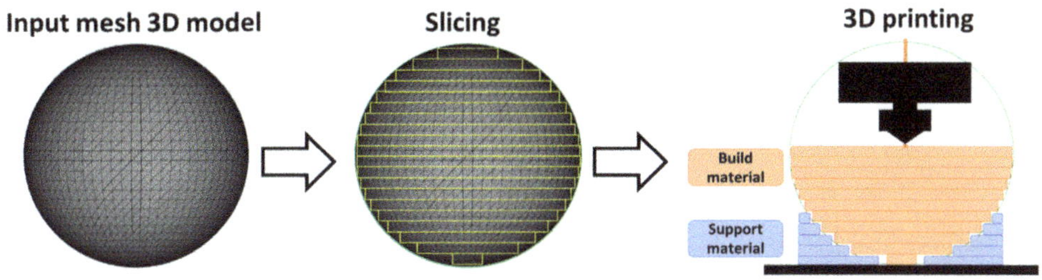

Figure 5.1. Partitioning of a simple 3D sphere model in slices during the slicing process.

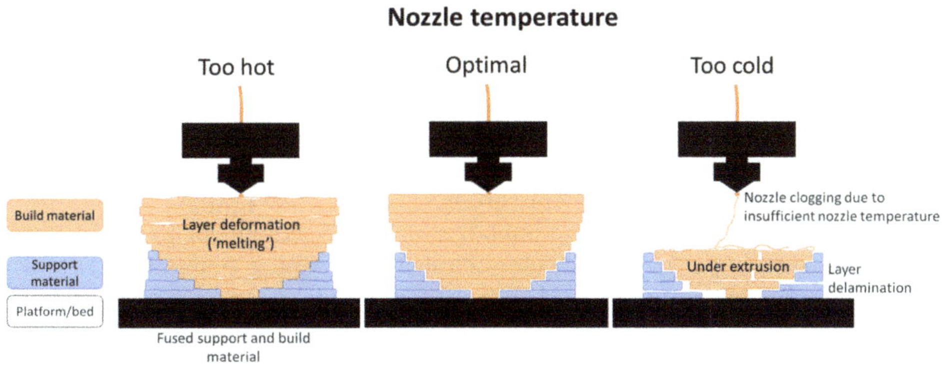

Figure 5.2. Impact of the nozzle's temperature on printing quality. On the left-hand side, high temperatures are shown deforming the layers of the print, while on the right-hand side low temperatures lead to clogging of the nozzle and insufficient extrusion and layer delamination.

The software used for slicing allows users to control 3D-printing parameters such as layer height (known as the 'z' resolution), printing speed, and the use of support structures. The overall slicing process is illustrated in figure 5.1. The model slicing process varies across different 3D-printing technologies, depending on the 3D-printing parameters that a user can control, although many parameters are common to multiple 3D-printing technologies.

5.2.2 Nozzle temperature

This controls the temperature of the nozzle head as the filament is extruded. The temperature is chosen to be reasonably above the melting point of the filament because low temperatures may lead to ineffective melting of the filament and blocking of the nozzle. The quality of the extrusion and adhesion between printed layers are sensitive to the nozzle's temperature. At higher temperatures, there is generally greater adhesion of the extruded filament, but this usually corresponds to a loss of printing resolution and can lead to deformation. Excessively high temperatures risk damaging the extruded filament. The effects of nozzle temperature selection for MEX or similar printing systems are illustrated in figure 5.2.

While most slicing software and filament manufacturers provide a recommended range of nozzle temperatures, this setting may need to be optimised to obtain high quality prints (see chapter 8).

5.2.3 Build plate temperature

The printing build plate (sometimes called the 'print bed') provides a surface to print the object, and is also responsible for creating adequate adhesion and to minimise warping of the base layer. Even below the melting point, 3D-printed materials remain malleable. It is important for the printed layers to remain above the material's heat deflection temperature to lay flat on the printer plate. Just like the nozzle's temperature, if the plate's temperature is set too high or too low, then it can create problems for the print. Too high a temperature will fuse the material too strongly to the build plate's surface, making it difficult to remove. If the initial layers are not adequately cooled prior to additional layers being printed, then this may cause an 'elephant's foot' problem. This may be mitigated through the use of a raft or through design adjustments, such as chamfering, if geometric accuracy is required. For low build plate temperatures, the adhesion to the plate may be insufficient, causing warping along the edges or total detachment from the plate—potentially producing a 'birds nest' (see chapter 8). Warping generally occurs when different components of the print cool at different rates, and can be mitigated by more even cooling (e.g. adding or removing an enclosure, modifying fan settings, or modifying plate and initial layer nozzle temperatures), or the use rafts or brims. The effects of build plate temperature selection for MEX or similar printing systems are illustrated in figure 5.3. These issues are unfortunately amplified when printing large objects with high infill percentages, such as bolus or phantoms.

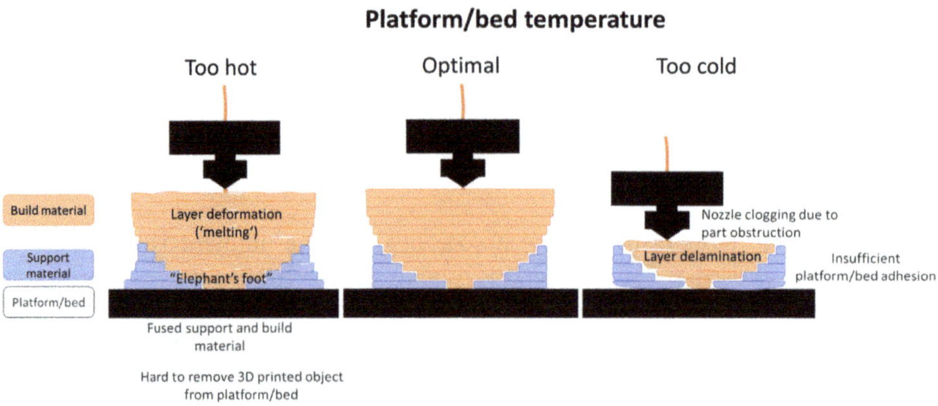

Figure 5.3. The effect of the build plate temperature is shown. On the left, the high plate temperature leads to layer deformation and a larger primary layer, referred to as an 'elephant's foot'. On the right, insufficient temperature leads to the warping of the bottom layer.

5.2.4 Printing speed

The total time required for 3D printing an object is a function of many variables, including the total movement of the nozzle (or laser for VPP, MJF (SLS) or SHL systems, see chapter 2), the speed of that movement, the total amount of material that must be deposited (or cured, sintered, etc), and the rate at which the material deposition (or curing, sintering, etc) occurs. These variables are each a product of others; for example, the total distance travelled depends on the number of layers, the bounding volume of the print in each layer, the infill pattern, and the use of scaffolding (which depend on the orientation of the print etc). For MEX printers, the rate at which material deposition occurs also depends on width of the extrusion, or nozzle diameter, and how quickly the motors are driving the filament.

Within most MEX print slicing software, there is typically a parameter referred to as 'print speed', defined in mm s^{-1}, which represents how quickly the print head moves. Slicing software may also provide additional control over printing speeds in specific circumstances. For example, the initial layers or external shells (sometimes called 'walls') can be printed more slowly than the internal volume, or the print head can be moved faster when not extruding. However, if this print speed parameter is set too high, then it can have a considerable impact on the quality of the printed object (e.g. leading to poor layer adhesion at the beginning of the print, under extrusion when the filament drive cannot keep up with print head motion, and errors in positioning due to sudden changes in print head momentum). The print speed also determines how long the material is heated within the nozzle: too slow, and there can be oozing and stringing; too long, and it can fail to extrude, leading to clogging.

Shorter print times are advantageous, not just due to clinical turnaround and machine utilisation but also because longer prints can be subject to other issues, such as warping due to uneven cooling, or heat creep in the print head may cause the filament to soften or melt further away from the nozzle, potentially causing clogs.

5.2.5 Layer height

The selection of layer height (also called layer thickness) will determine the print speed and impact the quality of the print in terms of surface smoothness and geometric precision. The minimum and maximum layer height will depend on the printer being used (e.g. VPP printers often use 0.05 mm as a default setting, whereas MEX printers may default to 0.2 mm). For MEX printers, it is generally recommended that a layer height between 25% and 75% of the nozzle diameter be used, which is typically 0.4 mm unless replaced. Nozzles can be purchased with diameters ranging from 0.2 to 1.2 mm, depending on the MEX printer, with larger nozzles being suitable for printing composite materials (which contain particulates that might clog or damage smaller nozzles).

Depending on the 3D model, print speed can scale near linearly with layer height (i.e. doubling the layer height will nearly halve the print time). Despite this speed cost, smaller layer heights can be advantageous because they produce higher-resolution prints. Smaller layer heights also inevitably reduce the amount of 'overhang' in each layer, because each layer will protrude less than the layer below.

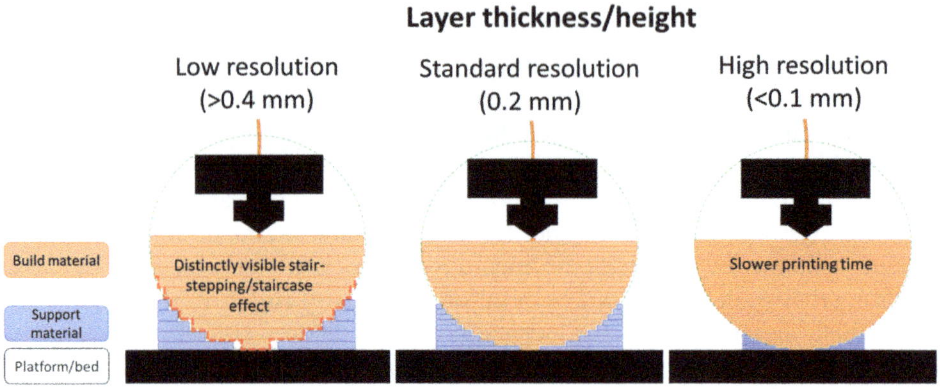

Figure 5.4. The effect of layer thickness. On the left-hand side, the increased thickness leads to a faster print with lower resolution. On the right-hand side, the reduced thickness leads to a slower print with higher resolution and a reduced need for supports.

The use of a smaller layer height can therefore reduce the amount of scaffolding required to support these protrusions, which may be significant, depending on the 3D model. The impact of layer height on a MEX printer is shown in figure 5.4.

5.2.6 Infill density and pattern

When printing with a MEX printer, the density of an object can be modified through variation of the infill properties, from 0% (hollow) to 100% (solid). This affects the mechanical and also radiological properties of the printed device. For example, Asfia *et al* [4] investigated the effects of different infill densities (15%, 30%, 50%, and 80%) in immobilisation devices on skin dose and mechanical properties of the 3D-printed samples, and concluded that by increasing the infill density, the relative surface dose increases. For 3D-printed immobilisation devices, this surface dose enhancement is undesirable. However, increasing infill density also improves the print's mechanical properties. Therefore, a balance is needed, and will depend on the size and orientation of the print.

For radiation therapy bolus, compensators, range-shifters, and brachytherapy applicators (described in chapter 9), the infill density will often be optimised to produce a density similar to that of water or tissue. This can be estimated based on the measured density of a printed sample, but is often characterised by 3D-printing test objects with different infill densities and imaging or by performing transmission measurements [5]. These infill densities may need to be optimised for specific combinations of slicing software, 3D printer, specific material (potentially even batch of material), and infill pattern. Similarly, infill densities may be modified to achieve a desired non-water-equivalent density; for example, to mimic lung tissue in a quality assurance phantom (as described in chapter 10).

As with infill density, the pattern with which infill can be printed can also be selected, and can affect the mechanical and radiological properties. Some examples are shown in figure 5.5. Asfia *et al* [6] found that line and grid patterns were the

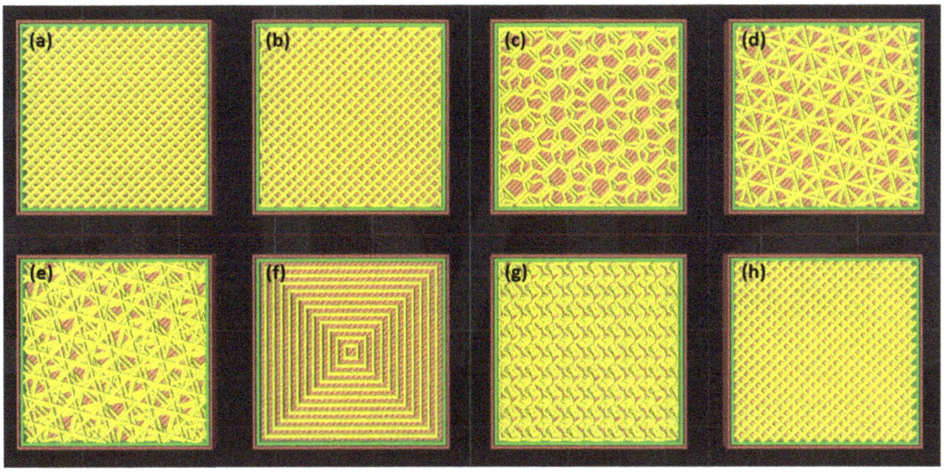

Figure 5.5. Infill patterns featured in ideaMaker software: (a) grid, (b) rectilinear, (c) honeycomb, (d) triangles, (e) cubic, (f) concentric, (g) gyroid, and (h) lines.

fastest to print, while cubic and gyroid patterns offered the greatest strength, and triangular and star patterns were a middle-ground. For flexible prints, a concentric infill pattern may be appropriate. For objects or devices where a homogeneous density is desired, regardless of angle of incidence of radiation, a gyroid infill is required [7]. An example of these patterns is shown in figure 5.5.

For VPP 3D printers, varying the effective infill is not simple because the internal structure will need to be defined prior to slicing and hollow cavities within a closed volume may require the addition of drainage holes for printing (see chapter 2).

5.2.7 Build orientation and support

The orientation of the 3D model relative to the print build plate will affect print time, and can determine the smoothness and accuracy of the print. The four principle considerations when selecting orientation should be:

1. Reducing the amount of overhangs and support material required for the print. Overhang, or the protrusion of each newly printed layer beyond the outer edge of the previous layer, can result in layer separation, geometric inaccuracies, or even print failure. When overhang exceeds 45°, support material or scaffolding may be required to prevent this overhang from sagging. These support materials will need to be removed from the printed object during post-processing, and affected surfaces may require substantial work to eliminate residual material and surface roughness. Minimising the need for support material can reduce material wastage and print time.
2. Consistent with the objective of reducing support material, it is important to ensure that the initial layer is suitable to maintain adhesion to the print build plate for the duration of the print. This is not necessarily the same as maximising the surface area of the initial layer because very large areas may be subject to uneven cooling and warping (see Craft and Howell [8] for

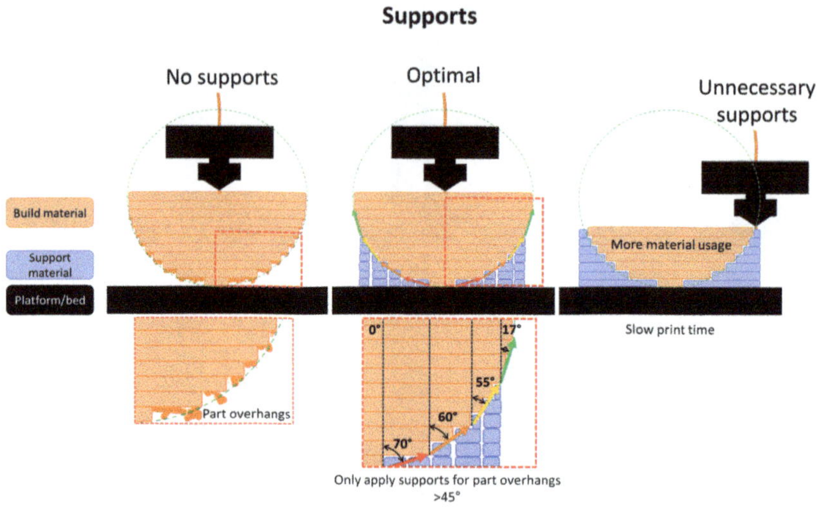

Figure 5.6. Effects of support structures are shown (insets show detail).

example, where thorax phantom slabs were prepared and printed in a sagittal orientation).

3. Care should be taken when determining the surfaces aligned along the vertical z-axis, perpendicular to the print build plate. The effect of layer lines, or the steps between layers, will be more noticeable for curved surfaces than for flat surfaces. Similarly, consideration should be given to the orientation of fine features, with dimensions comparable to that of the nozzle diameter and extrusion width, and the layer height. Depending on retraction settings, there may also be a 'z seam' or vertical line present on the printed object, resulting from the nozzle starting and finishing every layer at approximately the same x-y position and extruding a small amount of material while the step to each new layer is completed.
4. The mechanical and radiological properties of the printed object may vary with print orientation, depending on infill density, infill pattern, and layer height. Asfia *et al* [4] reported variations in downstream dose and ultimate tensile strength that depended on the orientation of internal air cavities with respect to the angle from which force was applied.

Figure 5.6 gives an example of a common situation where the need for support structures cannot be entirely avoided by rearranging the orientation of the object on the build plate, and illustrates the effects of inadequate and excessive supports.

5.2.8 Variable print parameters

MEX slicing software often allows the user to define different print parameters for different parts of the print; for example, varying infill densities for different regions (or different materials for multi-nozzle printers) allows the production of

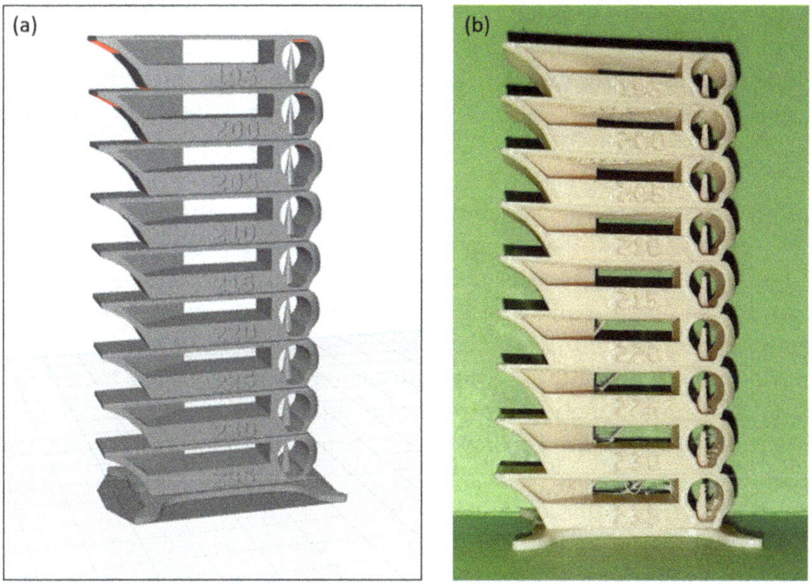

Figure 5.7. Temperature tower designed by Gaa [9] (a) 3D model of intended print and (b) completed object with each section printed at the temperature in °C indicated by surface markings, showing variable stringing and overhang failure with print temperature.

sophisticated anthropomorphic phantoms [10, 11], as described in chapter 10. This capability can be used to evaluate the effect of variations of printing parameters; for example, by printing a temperature tower, as shown in figure 5.7 [9].

5.3 Printing

After preparing the 3D model for printing as described throughout section 5.2, the resulting print instructions (whether g-code or an alternative) need to be loaded onto the printer. This may involve physical transfer by microSD, SD, or USB storage, or via the local network or cloud, depending on the hardware. Some printers can be directly controlled by connected computers, including inexpensive single-board computers [12], which may be able to be connected to internal networks, simplifying this data transfer. The amount of descriptive information or metadata included in print instruction files varies with the slicing software and hardware used. Unique, descriptive file names should be used wherever possible, with consideration for patient confidentiality requirements.

Before starting the print, care should be taken to ensure that the material loaded in the printer is correct and the material should be visually inspected for defects. Potential issues that can be identified at this stage include breaks or tangling of a filament, separation of resin components, and expiration of the material beyond shelf life. Many MEX filaments can degrade over time due to absorption of moisture, so air-tight storage of opened filaments should be considered (see chapter 6). For printers that do not automatically pause when material has run-out (e.g. through the use of an optical filament sensor), the amount of material loaded should be sufficient to complete the print. Depending on the implementation of the run-out pause, it may be desirable to estimate when the run-out will occur in order to replace the material quickly (e.g. if the nozzle remains at extrusion temperature while paused). Scales are useful for estimating the amount of material remaining on a spool or in a container.

Similarly, the printing surface should be inspected to ensure they are clean and smooth. 3D-print failures often occur near the beginning of a print; for example, as a result of poor adhesion between the first layer and the print build plate due poor levelling or nozzle height calibration or an unclean plate surface. For printers that are not being actively monitored, checking adhesion early can minimise wasted time and materials. Some printers have integrated web-cameras to allow remote monitoring, and this functionality can be added to existing systems through solutions such as OctoPrint [12]. For MEX printers, there are various techniques that can be used to improve adhesion, depending on the plate material, such as coating with hairspray, adhesive glue, or painter's tape.

The use of a schedule or log system to track use of the printer is recommended, including contact details for the job owner, clinical prioritisation, the amount of type of material to be used, and the start and estimated completion times. Where printing resources are limited, prints exceeding the length of the workday can be scheduled to run overnight on printers with demonstrably effective thermal regulation, improving printer utilisation. For printers where replacement of materials can be laborious or time-consuming (e.g. SLS printers), prints in the same materials should be scheduled sequentially.

A log system to report print failures or non-conforming devices, printing issues, and completion of preventative maintenance or repairs is also recommended. Staff responsible for the printing of radiation therapy devices or maintenance of the printers should be familiar with common 3D-printing issues (discussed in chapter 8) and be prepared to provide support as necessary. Workflows for print preparation, printing, and post-processing should be documented, and be made available to all relevant staff.

5.4 Post-processing

Post-processing in 3D printing refers to the steps that are taken after the 3D-printing process is complete. These steps may include removal of the print from the print build plate, removing support material, removing residual material such as uncured resin, smoothing rough surfaces, or adding a finish to the printed object. Some common techniques for post-processing 3D-printed objects include washing, sanding, painting, and polishing. Some 3D printers also have built-in post-processing features such as a heated build plate or chamber that can be used to improve the surface finish of the printed object. Post-processing is an important step in the 3D-printing process because it helps to improve the appearance and function of the final object.

The order with which post-processing will be completed varies with the printing technology and the user's preference. For example, for resin-based systems, the printed object could be rinsed or bathed in isopropyl alcohol for a short period of time, any supports should be physically removed, prior to final curing with light, avoiding the difficult removal of hardened supports. For powder-based printers, the first step might be removal (and recapture) of residual un-sintered powder, which may be brushed off, shaken off by mechanical vibration, or vacuumed, followed by blasting or tumbling to improve the surface finish, and finally the object can be given an application of cyanoacrylate.

Removal of objects from MEX printers is not always simple, depending on level of adhesion and print surface. Removal is often easier for MEX printers once the build plate has been left to cool to room temperature. Caution should be used when using brute force, due to risk of damage to the print or printer and the risk of personal injury. Flat blades or spatulas are often used, but again care is required to avoid damage to the printed device or print surface, or personal injury. Flexible build platforms are available for some MEX printers, which provide a build plate that can be easily removed from the printer and flexed or bent until the print detaches.

The removal of supports will depend on the material used. Some scaffolding material is dissolvable in water, or may be removed using a corrosive bath. Other materials may require manual removal using hand tools, such as pliers, flush cutters, deburring tools, dremels, or rotary grinders.

Depending on the material that the object is made of, surface smoothing or finishing may involve machining (e.g. for metal parts), abrading with a file, rasp, sandpaper, or a linishing machine, or using a chemical solution, such as vapour smoothing (e.g. the use of acetone with ABS plastics). If required, objects may be painted, powder coated, or bath dyed. Surface plating can be used to add strength or other properties such as conductivity.

Many of the tools used for removal of support material and smoothing have associated risks to the user, including the risk of direct injury, and the risk of exposure to fine particles. Care should be taken when using this equipment (see chapter 6).

Devices printed in multiple parts may be assembled using joints or keys (e.g. interlocking or snap-fit joints) or holes for fasteners added during the design stage, or more permanently with the use of glues such as cyanoacrylate for polymers.

Prior to any use of the printed object, storage and labelling must be considered to ensure safe use. Quality assurance should be completed as required (see chapters 7 and 8), and, where necessary, the object may be cleaned with an alcohol solution or sterilised (see chapter 9).

5.5 Key points

- Prior to 3D printing, a 3D model must be 'sliced' to produce instructions (usually g-code) for the 3D printer.
- The slicing process can determine the quality of the printed object or device, especially for MEX printing, due to the importance of printing parameters

such as nozzle temperature, build plate temperature, printing speed, layer height, infill density, infill pattern, and build orientation, as well as the variation of selected parameters during printing.

- Best results can be expected if optimal 3D-printing parameters (for a given 3D printer, material and printed device type) are identified prior to clinical use through a dedicated commissioning process (see chapter 8).
- Many 3D-printed radiation therapy devices require post-processing, including removal of supports. This post-processing requires the use of additional tools (see chapter 6).
- There are many steps involved in the preparation, printing,and post-processing of radiation therapy devices. Therefore, it is important that educational material and checklists be developed (see chapter 7), particularly if responsibility for printing of devices is distributed among staff members.

References

[1] Tino R, Yeo A, Leary M, Brandt M and Kron T 2019 A systematic review on 3D-printed imaging and dosimetry phantoms in radiation therapy *Technol. Cancer Res. Treat.* **18** 1533033819870208

[2] Rooney M K, Rosenberg D M, Braunstein S, Cunha A, Damato A L, Ehler E, Pawlicki T, Robar J, Tatebe K and Golden D W 2020 Three-dimensional printing in radiation oncology: a systematic review of the literature *J. Appl. Clin. Med. Phys.* **21** 15–26

[3] Albantow C E and Brown S J 2022 Three-dimensional printer use in the Australian and New Zealand radiation therapy setting *J. Med. Radiat. Sci.* **69** 492–501

[4] Asfia A, Deepak B, Novak J I, Rolfe B and Kron T 2020 Infill selection for 3D printed radiotherapy immobilisation devices *Biomed. Phys. Eng. Express* **6** 065014

[5] Dancewicz O, Sylvander S, Markwell T, Crowe S and Trapp J 2017 Radiological properties of 3D printed materials in kilovoltage and megavoltage photon beams *Phys. Med.* **38** 111–8

[6] Asfia A, Novak J I, Mohammed M I, Rolfe B and Kron T 2020 A review of 3D printed patient specific immobilisation devices in radiotherapy *Phys. Imaging Radiat. Oncol.* **13** 30–5

[7] Tino R, Leary M, Yeo A, Brandt M and Kron T 2019 Gyroid structures for 3D-printed heterogeneous radiotherapy phantoms *Phys. Med. Biol.* **64** 21NT05

[8] Craft D F and Howell R M 2017 Preparation and fabrication of a full-scale, sagittal-sliced, 3D-printed, patient-specific radiotherapy phantom *J. Appl. Clin. Med. Phys.* **18** 285–92

[9] Gaa Z 2017 *Smart Compact Temperature Calibration Tower* [Online] Available: https://www.thingiverse.com/thing:2729076 (visited on 05/30/2023)

[10] Tino R B, Yeo A U, Brandt M, Leary M and Kron T 2022 A customizable anthropomorphic phantom for dosimetric verification of 3D-printed lung, tissue, and bone density materials *Med. Phys.* **49** 52–69

[11] Kairn T, Zahrani M, Cassim N, Livingstone A, Charles P and Crowe S 2020 Quasi-simultaneous 3D printing of muscle-, lung-and bone-equivalent media: a proof-of-concept study *Phys. Eng. Sci. Med.* **43** 701–10

[12] Häußge G 2023 OctoPrint.org [Online] Available: https://octoprint.org/ (visited on 05/30/2023)

IOP Publishing

Chapter 6

Costs

Tanya Kairn, Rance Tino, Martin Leary and Adam Unjin Yeo

6.1 Introduction

A successful three-dimensional (3D)-printing service relies not only on the selection of the 3D-printing hardware, materials, and parameters (as described in chapters 2–5) but also on the management of factors that can be characterised as the costs of the service, including time and money as well as health hazards, physical space, and waste management. These factors should be assessed with reference to the costs of other pre-existing methods of fabrication or procurement. Some of the approaches to these considerations are described in this chapter, providing a foundation for the implementation of a well-informed 3D-printing quality management system, as described in chapter 7.

6.2 Money

Alongside the ability to easily produce patient-specific devices (see chapter 9) and other purpose-designed equipment (see chapter 10), the other major attraction of 3D printing for radiation therapy is the low monetary cost of the printed objects when compared to equivalent commercial alternatives. This is especially noticeable when it comes to fabricating measurement jigs and phantoms for radiation therapy system and treatment verification (see chapter 10). For example, consumer-grade material extrusion (MEX, aka FDM or FFF) printers (see chapter 2) costing no more than few hundred dollars, pounds, or euros, have been reported as producing phantom components and treatment accessories with a high degree of geometric precision and density fidelity using just tens of grams of plastic filament [1, 2].

To fabricate complicated anthropomorphic phantoms, sophisticated dual-nozzle printers can now be purchased for less than one third of the cost of in-house fabrication of just one phantom in 2011 [3], or less than a quarter of the cost of purchasing one anthropomorphic dosimetry phantom from a commercial supplier today. The setup costs of 3D-printing hardware can also be reduced through the use of open-source designs, components, and software.

doi:10.1088/978-0-7503-3907-0ch6

When estimating the monetary cost of a clinical 3D-printing service, it is important to take into account expenses beyond printer and filament purchases. Computer workstations, design software, and post-processing equipment may need to be purchased (see chapters 4 and 5), especially as the service expands. There will also be recurring purchases of consumables, such as cleaning and disinfectant chemicals, as well as personal protective equipment (PPE) (see chapter 9 and section 6.4).

The time needed to set up and operate a clinical 3D-printing service amounts to a direct cost in terms of staffing (see section 6.3). There may also be monetary costs related to the handling of health risks (see section 6.4), the physical space required to run the service (see section 6.5), and the handling of waste material produced by the service (see section 6.6).

As addressed further in section 6.3, the monetary costs of implementing the 3D-printing service should also be weighed against the clinical costs of not implementing the service.

6.3 Time

The reported times taken for 3D printing of radiation therapy devices can range from a few hours for small boluses [4], to 30 hours for full face positives used for shielding design (see chapter 9), and to many days for large complex phantoms that are printed in sections [5]. These fabrication times may seem long, especially when compared to the sometimes less than 30 minute moulding times required to manually form treatment devices using wax [6] or the negligible time needed to apply a generic sheet bolus. However, these 3D-printing times should be balanced against the improved treatment outcomes and patient comfort achievable using patient-specific 3D-printed devices [7–9].

For MEX printing, radiation therapy device fabrication times can be reduced by using increased nozzle sizes, layer heights, and print speeds, although the effects of these parameters on print quality should be carefully optimised (see chapter 5).

Additionally, it should be understood that long print runs may be completed with minimal staff oversight, including overnight, as long as the 3D printer has been carefully set up and commissioned (including verification of long print runs), has been shown to have good control against thermal runaway, and is kept in a well-ventilated space. With sufficient initial testing and thoughtful use of scripting and automation, the creation of 3D-printed radiation therapy devices may take less staff time, and definitely less patient time, than current alternatives such as moulding in wax, plaster, thermoplastic, and silicone (see chapter 9).

As efficient as the clinical 3D-printing service will eventually be, the initial setup of the service will inevitably have an appreciable cost in terms of staff time. Many hours must be invested by multi-disciplinary teams to identify the goals of the service, develop relevant workflows, and set up a quality management system (see chapter 7). For the best and safest results, time must also be invested in testing and commissioning hardware, software, and materials, and in developing a quality assurance system with appropriate quality control tests and tolerances (see chapter 8). In some jurisdictions and for some clinical applications, time may also need to be invested in seeking regulatory approval for the 3D-printing service (see chapter 7).

Given the manifold advantages of 3D printing for radiation therapy in terms of process efficiency, patient experience, and treatment effectiveness, it would be advisable for health services, trusts, and funding agencies to provide funding support to cover staff costs during the period of initial setup of increasing numbers of clinical 3D-printing services.

6.4 Health

It is well known that until they are cured and made inert, the photopolymer resins that are used in 3D-printing processes such as vat photopolymerisation (VPP) and multi-jet printing (see chapter 2) can cause irritation if skin contact occurs, and more serious damage if ingestion or eye contact occurs. The metallic and other powders used in processes such as binder jetting (BJT), direct energy deposition, selective layer sintering (SLS), and powder bed fusion (see chapter 2) also represent inhalation risks. It is important to follow proper safety protocols, such as wearing PPE and washing hands thoroughly after handling these materials, to minimise the risk of health issues. However, even the nominally stable plastics used in MEX printing can have detrimental effects on air quality, and therefore health, if not managed appropriately.

If 3D-printed radiation therapy accessories are to be fabricated within a hospital or clinical environment, then potential effects on air quality within the hospital or clinic must be considered as components of patient and staff safety. When considering the effects of 3D-printing systems on air quality, there are two key areas of concern: production of vapours and emission of airborne particles. Vapour production, including volatile organic compounds (VOCs), has been reported for VPP and MEX printers [10–14] and emissions of airborne particles, including nanoparticles or ultrafine particles (smaller than 100nm across), have also been reported for SLS and MEX printers [13, 15–18]. Figure 6.1 illustrates the production

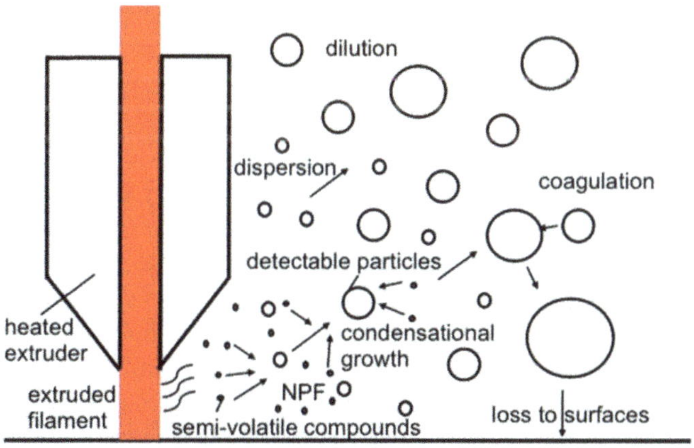

Figure 6.1. 'Schematic of particle formation, growth, and loss processes. NPF (new particle formation) is new particle formation resulting from nucleation of emitted semi-volatile vapours' [17]. Reprinted from [17], Copyright (2017), by permission of the publisher (Taylor & Francis Ltd, http://www.tandfonline.com).

of ultrafine particles and VOCs during MEX printing. Vapours may produce unpleasant odours, leading to discomfort [11, 14, 19], and exposure to VOCs in combination with ultrafine particles from 3D printers has been associated with the development of inflammatory and allergic responses [14, 15] and the exacerbation of respiratory issues such as asthma [15].

Given that the study of health effects from 3D printers is a comparatively new area of investigation, it has been recommended that 3D printers be used in ways that limit emissions, especially around vulnerable individuals [20]. Children have been specifically categorised as potentially vulnerable [17, 20], although this category should also include cancer patients and especially patients with pulmonary cancers or respiratory comorbidities.

For the health of patients and staff, care should be taken to ensure that rooms containing 3D printers are spacious and adequately ventilated [12, 14, 16, 19–21] (fume hoods may be required if space or ventilation is limited), and that large numbers of 3D printers are not compacted together [15, 21]. For MEX printers, the production of VOCs can be substantially reduced through the careful choice of printing filament. For example, in terms of VOC production, PLA is preferable to acrylonitrile butadiene styrene (ABS), but some PLA additives produce increased emission of ultrafine particles [13]. It may also be possible to minimise emissions through careful selection of 3D-printing hardware [17] and small changes to 3D-printing procedures [17, 18]. For example, several authors have noted that the rate of ultrafine particle emission from MEX printers is greatest at the start of the printing process [14, 17, 18], and Deng *et al* [18] have advised that particle emission can be reduced by heating the print head before loading the filament.

Regarding mechanical and electrical safety, it should be understood that each 3D-printer has been designed to print with certain types of materials under certain conditions, and therefore each printer should be operated within its own recommended working limits. Printer designs range from consumer-grade (hobbyist) machines with few inbuilt safety features, allowing direct exposure to moving parts and hot components, to high-quality commercial machines with interlocks and fume filtration. 3D-printing workflows typically include post-processing steps that may involve chemical baths or high pressure cleaners to remove support structures or harden the printed object, as well as power tools such as electric grinders or sanders for smoothing of printed surfaces. Consequently, users should be aware of the safety hazards associated with the use of their radiation therapy centre's 3D-printing and post-processing equipment, and safety training must be provided to all staff working with or near this equipment and access must be limited to trained staff only.

At the very least, to minimise the likelihood of patients straying into 3D-printing areas and to limit patient discomfort due to 3D-printer noise or odours, as well as potential patient exposure to 3D-printer air-borne emissions, 3D printers should be installed at a maximum distance from consultation rooms, nursing bays, imaging rooms, treatment rooms, and areas where patients spend large amounts of time, such as waiting areas.

Regulatory compliance and patient safety issues are addressed in chapters 7 and 9.

6.5 Space

Point-of-care manufacturing has great advantages in the radiation therapy context, due to the ability efficiently to design and create patient-specific devices that can then be adapted to changing patient anatomy or treatment needs. This is the rationale for the inclusion of a mould room or radiation therapy workshop in many hospital designs. However, modernisation of the point-of-care radiation therapy device fabrication service through the inclusion of a 3D-printing capability may require a re-evaluation of the amount of space required for the service.

The ideal radiation therapy 3D-printing room would be spacious, clean, well ventilated, sound proofed, and not crowded with 3D printers. Space should be allowed for access and clearance around 3D-printing and post-processing equipment, and ergonomic workstations and work benches should be provided for 3D-print preparation and post-processing (see chapter 5). Some examples of 3D-printing spaces in radiation therapy centres are shown in figure 6.2.

A range of storage options should be available, including cupboards or drawers to safely house post-processing equipment when not in use, as well as shelves accommodating dry boxes and air-tight storage for solid printing supplies, such as MEX filaments.

Space must be allotted for safe and secure storage of chemicals, including printing chemicals, and cleaning or sterilisation chemicals, with allowance for separate storage of hazardous, corrosive, or flammable chemicals, as well as light-tight storage of photopolymers. Fume hoods, extraction fans, or air filters may also be needed, depending on the types of chemicals, powders, and work to be done in the room.

Visible and well-signposted storage must be used for any necessary PPE, such as safety glasses, masks, gloves, and hearing protection. Fire extinguishers, blankets, or related equipment must also be accessible and clearly visible.

Electrical isolation may be advisable in addition to surge protection, especially in regions where the electricity supply may be erratic. Explosion-proof screens should also be available if any 3D printing of metals is to be undertaken using 3D-printing systems that do not have built-in safety screens.

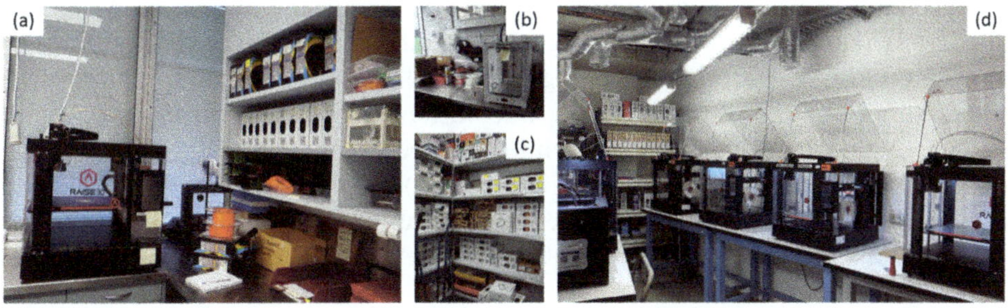

Figure 6.2. Examples of 3D-printing spaces in radiation therapy centres: (a) and (b) small 3D-printing areas in pre-existing mould rooms and workshops; (c) a small 3D-printing storage area; and (d) a dedicated 3D-printing room.

A dedicated 3D-printing room may not be justifiable for a small 3D-printing service with (for example) only one or two MEX printers. In these cases, a 3D-printing area should be specifically assigned and resourced after evaluation of relevant health hazards (see section 6.4) and with consideration for the ventilation, workspace, and storage space requirements described above.

6.6 Waste

As noted in chapter 3, 3D printing for radiation therapy is dominated by plastics. Widespread adoption of 3D printing for radiation therapy may have mixed effects on the global burden of plastic waste: on the one hand, plastic waste from industrial processes may be reduced if point-of-care manufacturing takes over some of the work previously done by centralised commercial companies; while on the other hand radiation therapy treatment devices such as bolus may be increasingly made from plastic when they were previously moulded from reusable wax or thermoplastic. Commissioning of clinical 3D-printing services inevitably involves the production of substantial amounts of waste via test prints and failures on the way to identifying optimal printing parameters (see chapter 8).

The production of many 3D-printing plastics from oil or petroleum [22], and the energy usage involved in producing, shipping, and using these materials [23, 24], implicates 3D printing as a contributor of atmospheric CO_2, especially in locations where a large proportion of the electricity supply is generated by burning fossil fuels. Again, manufacturing at the point-of-care will mitigate the CO_2 cost of shipping individual devices, but the CO_2 cost of transporting raw 3D-printing materials will remain an area of concern unless (or until) a substantial increase in point-of-care recycling is achieved.

The environmental cost of widespread 3D printing of plastic radiation therapy devices may be minimised through optimisation of 3D-printing methods, careful selection of materials, and (ideally) recycling 3D-printer waste at the point of production. First, it must be understood that in the radiation therapy context, 3D-printing waste is produced directly in the form of print supports as well as failed or otherwise unusable prints, and also indirectly with the eventual disposal of every patient-specific device as soon as the clinical need for that device has passed.

Given that a substantial amount of 3D-printed material will become waste either immediately (supports) or after use (patient-specific devices), it is very important to minimise the use of material and energy during the initial 3D-printing process. 3D-printed supports can be minimised and often eliminated through thoughtful planning of the orientation of the printed device on the printing build plate and the use of reduced layer heights (see chapter 5). The likelihood of material and energy wastage due to printing failures can be minimised through identification of optimal print parameters and orientations at the service commissioning stage (see chapter 8).

During MEX printing, Ma *et al* [24] have shown that the energy consumption of the heated nozzle and heated build plate dominates the 3D printer's energy usage, with the stepper motor, microcontroller, and fan making only small contributions.

For this reason, it is clear that a fast print is a more energy-efficient print than a slow print for MEX systems. Accordingly, Ma *et al* [24] found that energy consumption was decreased when printing was sped up by decreasing the infill density, and also increasing the layer height and printing speed. Obviously, when the print detail and print precision of 3D-printing devices for radiation therapy use are detrimentally affected by increasing layer heights and printing speeds (see chapter 5), these parameters should be optimised to avoid unnecessary energy use while maintaining suitable print quality. Similarly, infill density in radiation therapy applications is usually determined more by radiological effects than printing efficiency (see chapter 10), but efforts could be made to minimise infill density in appropriate contexts. Notably, Faludi *et al* [23] have highlighted the effect of printer idling on overall energy usage, with various types of 3D printers showing considerable inefficiency when left idling between completing just one print per week. Evidently, 3D printers should be switched off when not in use.

The environmental impact of 3D printing may also be reduced through careful selection and handling of printing materials. On the one hand, the use of hazardous and non-recyclable materials may be unavoidable for some applications (see chapter 3), and in these situations it is important that the materials are handled with care, and disposed of safely and in compliance with local regulations. On the other hand, there are many 3D-printing applications where one material might be substituted for another less environmentally-impactful material, without detriment. For example, MEX filaments supplied on cardboard spools should be preferred over the same filaments supplied on plastic spools. Additionally, many authors recommend PLA and ABS filaments for equivalent purposes (see chapter 9), and yet ABS is less advantageous for both safety reasons (see section 6.4) and for its increased environmental impact [23, 25].

PLA is considered to be a comparatively environmentally-friendly MEX filament due to its being derived from plant matter rather than from fossil fuel production. However, Kumar *et al* [25] have highlighted the increased water usage required for polylactic acid (PLA) filament production and suggested that polyethylene terephthalate glycol (PETG) filament may be less environmentally costly than PLA, although both filaments were observed to be preferable to ABS [25].

Additionally, despite its marketing and while it is still more biodegradable than many other plastics, PLA is technically only biodegradable in select industrial compositing facilities, under controlled temperature and humidity conditions, and with exposure to relevant microorganisms. Under these conditions, PLA will biodegrade within a period of days to weeks. PLA's degradation characteristics vary when released into the environment, ranging from medium in soils to low in waters [26]. In addition, PLA either does not degrade entirely or does not degrade quickly enough to avoid accumulation in oceans [27, 28]. In the ocean, PLA is similar to other 3D printing filaments such as ABS, in that it can form microplastics, which have been shown to have measurable health effects on organisms from plankton to fish [27, 28]. Ali *et al*'s [27] review of the literature on PLA microplastics indicated that these contaminants are no less detrimental to aquatic biota than microplastics from conventional petroleum-based plastics.

Given the substantial environmental costs of producing new 3D-printing plastics and releasing waste 3D-printing plastics into the environment, a means to recycle these materials in-house without incurring additional energy costs from transport to centralised facilities [29] is highly desirable. Fortunately, growing numbers of small, consumer-grade, commercial, and open-access plastic MEX filament recycling units are becoming available [29, 30]. After sorting, used 3D-printing plastics can be shredded before being melted and extruded back into filament form [30]. For PLA filament, repeating this melting and recycling process up to five times has been shown to produce minimal degradation of its mechanical properties [30], and observed changes in material viscosity and polymer chain length (molecular weight) can be avoided through the addition of a proportion of fresh PLA pellets to the waste PLA mix prior to extrusion [31].

Clearly, the requirements for extra equipment and space, the need to sort and store each different type of used printing material, plus the need to re-melt the filament materials for extrusion into filament (another opportunity for VOC release) mean that recycling of 3D-printed plastics in a clinical environment may be too great a challenge for some radiation therapy centres, despite the monetary cost savings achievable when recycling replaces new filament purchases [29].

In such situations, every effort should be made to limit the production of 3D-printing waste as much as possible. Lu *et al* [32] have suggested that an advantage of plastic printing using the SLS method is that printing supports are not needed, and therefore material not used in one print can be immediately used in the next print, leading to an almost 100% material utilisation rate. Aldawood *et al* [33] have proposed creating radiation therapy bolus by 3D printing thin external walls and then filling with water, using 80% less material than printing in solid plastic. It would also be advisable to preferentially select recycled commercial filament, where available, if in-house recycling cannot be achieved.

6.7 Key points

- The monetary cost of 3D printing many items of radiation therapy equipment, especially dosimetry phantoms, can be substantially less than the purchase cost of commercial equivalents.
- The use of open-source designs, printer components, and software, in combination with dramatically increased use of locally recycled printing materials, have the potential to reduce the monetary costs of 3D printing even further.
- The time-costs of 3D printing and the production of excess waste can be minimised through careful optimisation of the printing parameters (see chapter 5).
- The health-costs of 3D printing can be minimised though the selection of more benign 3D-printing techniques (e.g. MEX instead of BJT) and materials (e.g. PLA instead of ABS), and by careful planning and arrangement of the 3D-printing workspace.

- The environmental impact of 3D printing can be minimised through careful optimisation of print settings to avoid wastage (see chapters 5 and 8) and by the introduction of point-of-care plastic recycling as an important aspect of point-of-care print creation.
- The costs of 3D printing may be balanced by the improved patient experience and treatment outcomes that are achievable using patient-specific 3D-printed devices (see chapter 9).

References

[1] Cleland S *et al* 2022 Development of a customisable 3D-printed intra-oral stent for head-and-neck radiotherapy *Tech. Innov. Patient Support Radiat. Oncol.* **23** 1–7

[2] Canters R A, Lips I M, Wendling M, Kusters M, van Zeeland M, Gerritsen R M, Poortmans P and Verhoef C G 2016 Clinical implementation of 3D printing in the construction of patient specific bolus for electron beam radiotherapy for non-melanoma skin cancer *Radiother. Oncol.* **121** 148–53

[3] Harrison K, Ebert M, Kron T, Howlett S, Cornes D, Hamilton C and Denham J 2011 Design, manufacture, and evaluation of an anthropomorphic pelvic phantom purpose-built for radiotherapy dosimetric intercomparison *Med. Phys.* **38** 5330–7

[4] Zhao Y, Moran K, Yewondwossen M, Allan J, Clarke S, Rajaraman M, Wilke D, Joseph P and Robar J L 2017 Clinical applications of 3-dimensional printing in radiation therapy *Med. Dosim.* **42** 150–5

[5] How S, Banjade D, Crowe S, Dillon G and Skimmings A 2023 An anthropomorphic 3D printed inhomogeneity thorax phantom slab for SBRT commissioning and quality assurance *Phys. Eng. Sci. Med.* **46** 575–83

[6] Cleland S *et al* 2021 Dosimetric evaluation of a patient-specific 3D-printed oral positioning stent for head-and-neck radiotherapy *Phys. Eng. Sci. Med.* **44** 887–99

[7] Zhang Y, Huang Y, Ding S, Liang J, Kuang J, Mao Q, Ying W, Shu Y, Li J and Jiang C 2022 A clinical trial to compare a 3D-printed bolus with a conventional bolus with the aim of reducing cardiopulmonary exposure in postmastectomy patients with volumetric modulated arc therapy *Cancer Med.* **11** 1037–47

[8] Elmahmoud Z, Gunther J R and Christopherson K 2023 A simple solution to create a custom scalp-sparing helmet to prevent alopecia in patients undergoing total skin electron beam therapy for cutaneous T cell lymphoma *Clin. Transl. Radiat. Oncol.* **38** 53–6

[9] Kairn T, Chan P, Chua B, Cleland S, Dawes J, Kenny L, Lin C Y, McDowall W R, Poroa T and Crowe S 2023 Magnetic resonance imaging assessment of the suitability and consistency of radiotherapy treatment positioning achieved using intra-oral stents *IFMBE Proc.* (in press)

[10] Randolph S A 2018 3D printing: what are the hazards? *Workplace Health Saf* **66** 164–4

[11] Gu J, Wensing M, Uhde E and Salthammer T 2019 Characterization of particulate and gaseous pollutants emitted during operation of a desktop 3D printer *Environ. Int.* **123** 476–85

[12] Wojtya S, Klama P and Baran T 2017 Is 3D printing safe? analysis of the thermal treatment of thermoplastics: ABS, PLA, PET, and nylon *J. Occup. Environ. Hygiene* **14** D80–5

[13] Floyd E L, Wang J and Regens J L 2017 Fume emissions from a low-cost 3-D printer with various filaments *J. Occup. Environ. Hygiene* **14** 523–33

[14] Gümperlein I, Fischer E, Dietrich-Gümperlein G, Karrasch S, Nowak D, Jörres R A and Schierl R 2018 Acute health effects of desktop 3D printing (fused deposition modeling) using

acrylonitrile butadiene styrene and polylactic acid materials: an experimental exposure study in human volunteers *Indoor Air* **28** 611–23
[15] House R, Rajaram N and Tarlo S M 2017 Case report of asthma associated with 3D printing *Occup. Med.* **67** 652–4
[16] Stabile L, Scungio M, Buonanno G, Arpino F and Ficco G 2016 Airborne particle emission of a commercial 3D printer: the effect of filament material and printing temperature *Indoor Air* **27** 398–408
[17] Zhang Q, Wong J P S, Davis A Y, Black M S and Weber R J 2017 Characterization of particle emissions from consumer fused deposition modeling 3D printers *Aerosol Sci. Technol.* **51** 1275–86
[18] Deng Y, Cao S-J, Chen A and Guo Y 2016 The impact of manufacturing parameters on submicron particle emissions from a desktop 3D printer in the perspective of emission reduction *Build. Environ.* **104** 311–9
[19] Davis A Y, Zhang Q, Wong J P, Weber R J and Black M S 2019 Characterization of volatile organic compound emissions from consumer level material extrusion 3D printers *Build. Environ.* **160** 106209
[20] Manoj A, Bhuyan M, Banik S R and Sankar M R 2021 Review on particle emissions during fused deposition modeling of acrylonitrile butadiene styrene and polylactic acid polymers *Mater. Today Proc.* **44** 1375–83
[21] Byrley P, George B J, Boyes W K and Rogers K 2019 Particle emissions from fused deposition modeling 3D printers: evaluation and meta-analysis *Sci. Total Environ.* **655** 395–407
[22] Feeley S R, Wijnen B and Pearce J M 2014 Evaluation of potential fair trade standards for an ethical 3-D printing filament *J. Sustain. Dev.* **7** 5
[23] Faludi J, Hu Z, Alrashed S, Braunholz C, Kaul S and Kassaye L 2015 Does material choice drive sustainability of 3D printing? *Int. J. Mech. Aerosp. Ind. Mechatronics Eng.* **9** 144–51
[24] Ma H, Zhang Y, Jiao Z, Yang W, He X, Xie G and Li H 2021 Comprehensive assessment of the environmental impact of fused filament fabricationproducts produced under various performance requirements *J. Inst. Eng. (India)* C **102** 59–73
[25] Kumar R, Sharma H, Saran C, Tripathy T S, Sangwan K S and Herrmann C 2022 A comparative study on the life cycle assessment of a 3D printed product with PLA, ABS & PETG materials *Procedia CIRP* **107** 15–20
[26] Haider T P, Völker C, Kramm J, Landfester K and Wurm F R 2019 Plastics of the future? The impact of biodegradable polymers on the environment and on society *Angew. Chem. Int. Ed.* **58** 50–62
[27] Ali W, Ali H, Gillani S, Zinck P and Souissi S 2023 Polylactic acid synthesis, biodegradability, conversion to microplastics and toxicity: a review *Environ. Chem. Lett.* **21** 1761–86
[28] Ribba L, Lopretti M, de Oca-Vásquez G M, Batista D, Goyanes S and Vega-Baudrit J R 2022 Biodegradable plastics in aquatic ecosystems: latest findings, research gaps, and recommendations *Environ. Res. Lett.* **17** 033003
[29] Kreiger M A, Mulder M, Glover A G and Pearce J M 2014 Life cycle analysis of distributed recycling of post-consumer high density polyethylene for 3-D printing filament *J. Cleaner Prod.* **70** 90–6
[30] Sanchez F A C, Lanza S, Boudaoud H, Hoppe S and Camargo M 2015 Polymer recycling and additive manufacturing in an open source context: optimization of processes and

methods *2015 Int. Solid Freeform Fabrication Symp.* (Austin, TX: University of Texas at Austin)
[31] Zhao P, Rao C, Gu F, Sharmin N and Fu J 2018 Close-looped recycling of polylactic acid used in 3D printing: an experimental investigation and life cycle assessment *J. Cleaner Prod.* **197** 1046–55
[32] Lu Y, Song J, Yao X, An M, Shi Q and Huang X 2021 3D printing polymer-based bolus used for radiotherapy *Int. J. Bioprint.* **7** 414
[33] Aldawood F K, Chang S X and Desai S 2020 Design and manufacture of a high precision personalized electron bolus device for radiation therapy *Med. Dev. Sens.* **3** e10077

Chapter 7

Quality management

Emily Simpson-Page, Deepak Basaula and Scott B Crowe

7.1 Introduction

Three dimensional (3D) printing capabilities offer a lot of value to a radiation therapy centre, allowing the manufacture of various treatment devices (see chapter 9) and radiation therapy treatment verification tools (see chapter 10) at point-of-care, which can improve the quality of care in the application of radiation in medicine. When considering whether 3D printing would be a useful tool in a particular radiation therapy centre, one must discern if 3D printing will allow for the accurate design of desired products in an efficient, safe, and cost-effective manner, more so than commercial alternatives, including off-the-shelf products or outsourcing of the 3D-printing service. Setting up a 3D-printing service in a radiation therapy centre requires assessment of three fundamental questions to ensure that the work required to appropriately establish the service is worthwhile:

1. Can the desired devices be safely and reliably manufactured using 3D printing?
2. Are the desired devices unable to be acquired by other means more easily at a lower cost? Or, would 3D printing allow more efficient or cost-effective production?
3. Does the radiation therapy centre have appropriate resources to support and maintain a 3D-printing service, in compliance with any applicable regulatory requirements?

If the answer is yes to all of these questions, then one can conclude that working to establish a 3D-printing service will be worthwhile. However, simply purchasing a 3D printer, consumable materials, and learning to design and print radiation therapy devices is not enough to achieve this aim. Responsibilities and processes must be established for the success of a 3D-printing service to be realised. Managing a 3D-printing service requires a meticulous approach. This is particularly so when initially setting up and rolling out the service, or when producing medical devices

doi:10.1088/978-0-7503-3907-0ch7

with risks of adversely affecting patient outcomes and which may be subject to regulatory requirements. Having high-quality processes in radiation therapy contributes to quality results for patients, and this is no different in the space of 3D printing for radiation therapy. The guidance summarised in this chapter is intended to ensure a robust, safe, and high-quality approach to 3D-printing implementation and integration in radiation therapy.

7.2 Quality management systems

The consistent provision of quality products (or services) is achieved through a quality management system (QMS), which ensures that the manufactured devices are of a high quality through the definition of organisational structure and responsibilities, the management of resources and risks, and the development and documentation of processes and procedures to ensure quality. Quality in this context means that the manufactured devices meet all applicable requirements, including regulatory standards, initial design specifications, and (ultimately) that the device can be used safely and effectively for its intended purpose. The principles of QMSs are described by the ISO 9000 [1] and ISO 9001 [2]. The ISO 13 485 standard [3] describes a QMS for the production of medical devices specifically.

The ISO recommends a process approach to quality management—quality is achieved by developing processes that consistently produce results that meet the desired standards. These processes need to be documented, managed, supported by resources, developed to produce results that conform with desired standards, and monitored on an ongoing basis. The implementation can be considered across three levels: a management level, a process level, and a device specific level as, shown in figure 7.1.

At the highest level, the requirements of the 3D-printing service may be specified in a quality manual or plan, analogous to a radiation management or radiation safety plan in the field of radiation protection. This should outline the approaches taken to managing a 3D-printing service; for example, resourcing, staff training, how new devices are implemented, key performance indicators, and how incidents

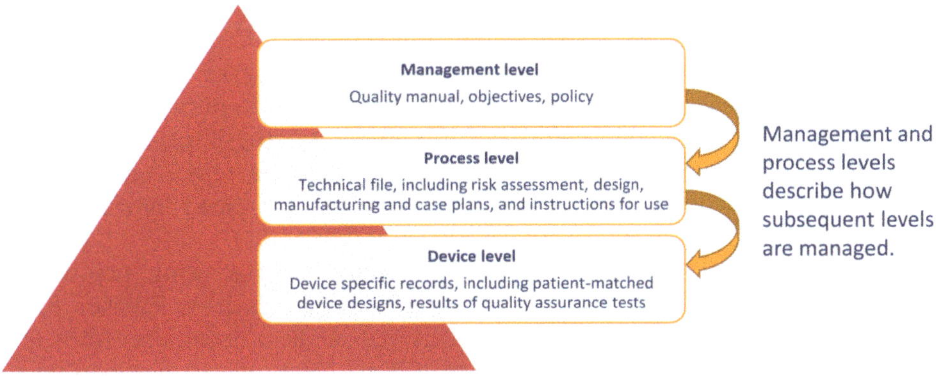

Figure 7.1. Levels of quality management.

are managed. Ensuring quality requires leadership and commitment from an organisation, with a clear communication of expectations of quality. Per ISO 9001 [2] and ISO 13 485 [3], this commitment should be expressed at the highest level of management in a quality policy and quality objectives, and should address the importance of the user and regulatory requirements. The quality policy might state, for example, that the organisation is committed to manufacturing devices that address the clinical needs of medical professionals in compliance with regulatory requirements. The quality objectives might reflect broader organisational objectives (e.g. commitment to the delivery of cost-effective high-quality patient care or the development of a culture of research and development) and should be measurable in some way.

These commitments must come with a clearly defined scope for the QMS and delegation of management responsibility to specific personnel to ensure that the QMS reflects the quality policy and objectives. These personnel (analogous to radiation safety officers in the context of radiation protection) will be responsible (for example) for ensuring that processes are appropriately documented and followed, that the organisation is adequately resourced, that risk is suitably managed, and that all regulatory requirements are met. For the provision of medical devices within a clinical radiation therapy centre, this delegation should be informed by consultation with a multidisciplinary team and with consideration of staff workload (discussed in section 7.5). Management arrangements should be reviewed periodically or when prompted by organisational or regulatory changes, and input should be sought concerning ongoing suitability and effectiveness (e.g. feedback from stakeholders or process monitoring data).

The process level captures the processes and documentation involved in the manufacturing and supply of specific types of devices. An example process, the production of a patient-specific radiation therapy bolus, is mapped in figure 7.2, capturing request, design, manufacturing, quality testing, and preparation for use. At the lowest level are the device specific records, such as the stereolithography (STL) files produced or the documentation of quality testing results.

For manufacturers of medical devices, quality management is especially important given the risks associated with the use of products that do not conform to clinical requirements or safety standards. The definition of a medical device varies with regulatory environments, but generally refers to any device intended for use in the diagnosis, treatment, mitigation, or prevention of disease. This includes the treatment devices described in chapter 9. Compliance with statutory and regulatory requirements for medical devices generally necessitates the implementation of a formal QMS. The scope and comprehensiveness of the QMS and degree of regulatory control will be commensurate with the risk presented by the manufactured device.

Implementation of a QMS ensures that good manufacturing practices are maintained for the organisation producing 3D-printed devices. Further to this, adherence to the QMS ensures that manufactured medical devices conform to the essential principles of safety and performance (i.e. that the device performs as intended, is safe to use, and that the benefits of its use outweigh the risks). A QMS

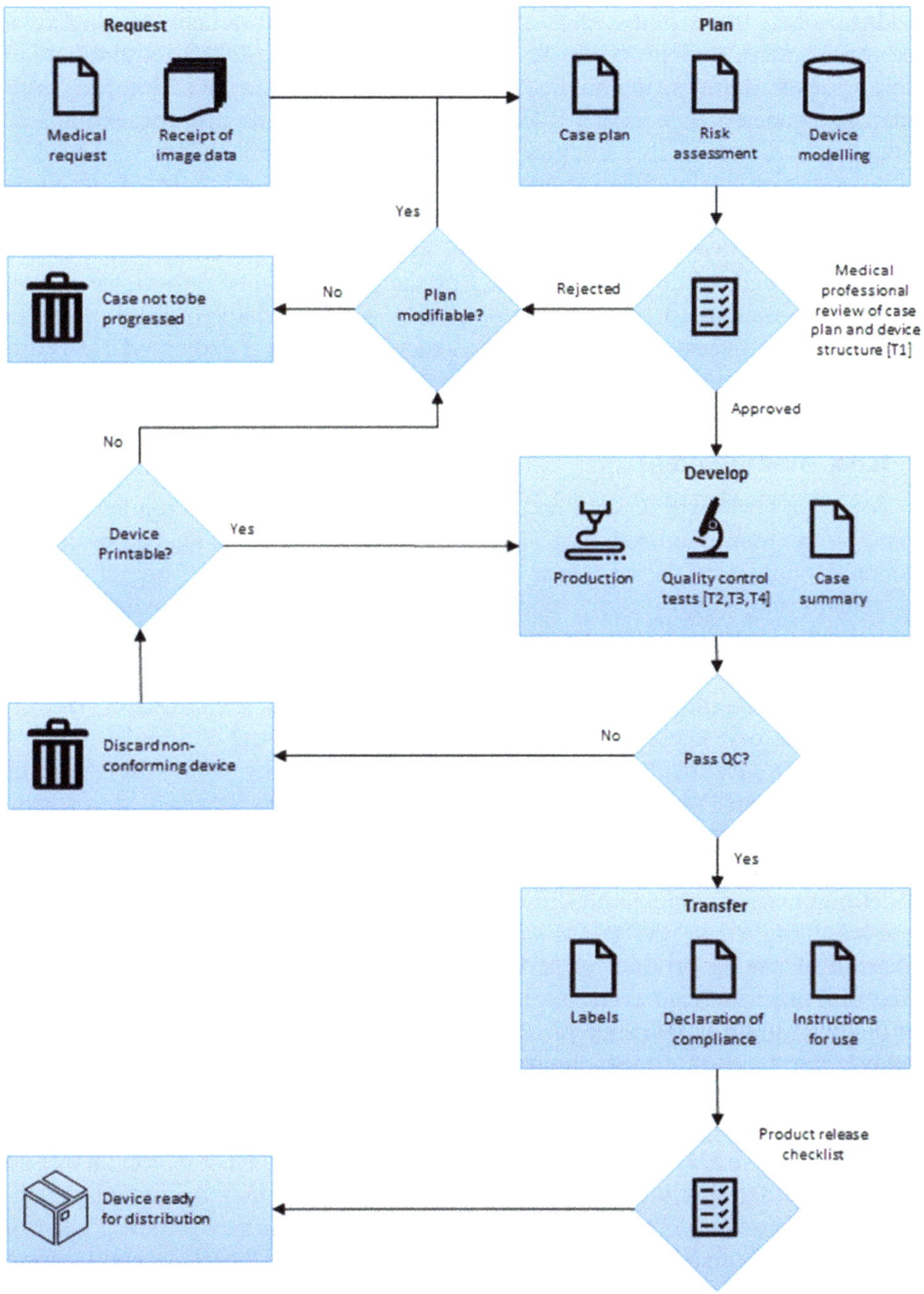

Figure 7.2. Example process map for the production of a radiation therapy bolus.

provides process traceability, allowing process failures and nonconforming devices to be investigated, and processes to be improved. Specific QMS requirements for organisations manufacturing medical devices are provided in ISO 13 485 [3] and by regulatory agencies. The ISO 13 485 [3] recommendations are contextualised in subsequent sections of this chapter for organisations implementing 3D-printing workflows to produce medical devices or devices with the potential to cause harm if quality is not managed (e.g. quality assurance tools), and the reader is encouraged to investigate their own local statutory and regulatory requirements prior to the production of medical devices in-house. For the production of non-medical devices without the potential to cause harm if quality is not managed, a comprehensive QMS will not be necessary, though record keeping (e.g. in a dedicated 3D-printing logbook) is good scientific practice.

7.3 Risk management

The scope of a QMS will depend on the risks associated with the intended use of the devices to be manufactured and supplied. As such, risk management must be considered once it is decided that a 3D-printing service is to be set up. Risk management is an integral facet of any application of radiation in medicine, but this is particularly important when implementing customisable and disruptive technology. Globally, risk management guidance and expectations are provided by institutions such as the ISO, International Electrotechnical Commission (IEC) and relevant regulatory agencies (e.g. the US Food and Drug Administration, the Japanese Pharmaceuticals and Medical Devices Agency, and the Australian Therapeutic Goods Administration).

The risk management process will include the specification of intended use of the device (with reference to the Global Medical Device Nomenclature (GMDN) code and definition where applicable), the identification of hazards associated with the device resulting from design or manufacturing failures or incorrect use, the estimation of risk (a product of severity and probability of harm), an assessment of acceptability of those risks, and (where possible) the implementation of risk controls (e.g. quality assurance practices). This process should be carried out by a suitably experienced multidisciplinary team and documented for future reference. If the overall benefits of the devices outweigh the risks, then those devices can be appropriately manufactured. The Report of the American Association of Physicists in Medicine (AAPM) Task Group 100 [4] provides advice on conducting risk assessments in a radiation medicine environment. Regulatory documents, such as the essential principles checklist defined by the Global Harmonisation Task Force and adopted by some regulatory agencies (e.g. the Australian Therapeutic Goods Administration) can help guide the risk assessment process. For example, the essential principles are used to assess a device's efficacy and safety through a list of rules, or guidelines, that should be checked for any given medical device. Local regulations may outline specific requirements depending on location of the manufacturing. As an example, figure 7.3 outlines a summary of the essential principles provided by the International Medical Device Regulators Forum.

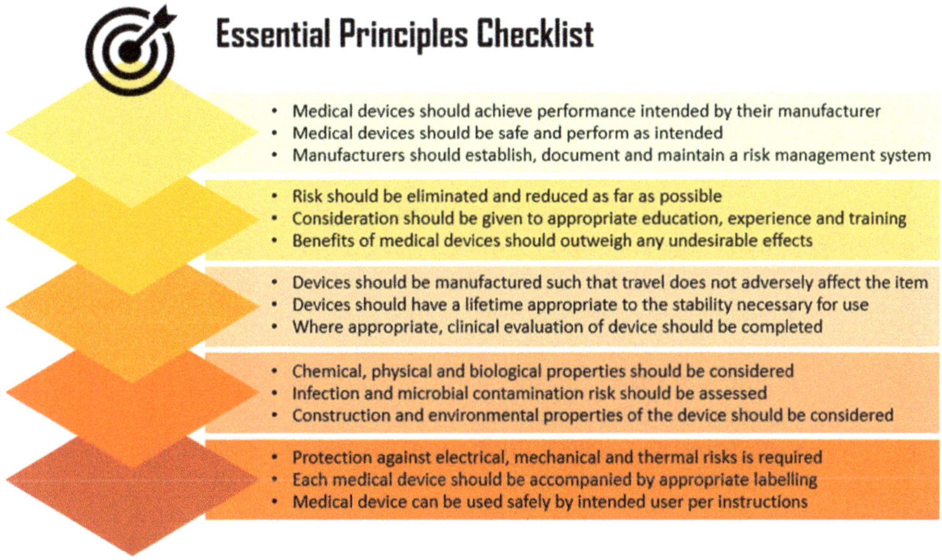

Figure 7.3. Summary of essential principles from the International Medical Device Regulators Forum.

The risk assessment determines the scope of the QMS, and forms the basis of the development of the processes and documentation that constitute the QMS itself. For example, the risk of incorrect use of a medical device should be identified during a risk assessment. This risk can be managed through the development of instructions for use for the device, which need to be accessible and understood by those who will use the device. To ensure that the instructions are accessible, the quality manual may describe a system for document control (e.g. networked file storage or an intranet page). To ensure that the instructions are understood by the intended users, the quality manual may describe a multidisciplinary review process for this documentation. Any instructions for use that are developed to mitigate the risks associated with misuse of the device will also be part of the QMS.

For medical devices subject to regulatory oversight, the degree of risk associated with the device is captured in a risk or device classification. For example, in Australia, the Therapeutic Goods Administration provides five outlines of risk classification from low to high. Classification can be determined by answering questions concerning the intended use of the device, how invasive the device will be, and where and how long it will be in use. Generally, one should consult their national guidelines on the production of medical devices and ensure that appropriate steps are taken to comply with any local, national, and international regulations that are applicable.

7.4 Documentation requirements

A comprehensive QMS will include a substantial amount of documentation, including descriptions of processes or workflows, standard operating procedures, assignment of staff responsibilities, specifications for devices (or types of devices), and manufacturing records. For medical devices, the production and maintenance of this documentation may be mandatory to achieve regulatory compliance. For both medical and non-medical devices, documentation and record-keeping is an integral part of maintaining quality.

ISO 9001 [2] and ISO 13 485 [3] recommend the development of a quality manual, which details the scope of the QMS and includes or references process documentation (see figure 7.1). A quality manual should be a high level document that can be referenced by staff or auditors, analogous to an internal radiation management or safety plan. The structure of this quality manual can reflect the structure of ISO 13 485 [3] and may specifically address how each clause of the standard is addressed within the organisation, explicitly stating, for example:

- What are the objectives of the service? And, how will the organisational leadership support those objectives?
- Who is responsible, accountable, and consulted regarding 3D-printing processes? And, what training or credentialing is provided?
- What sorts of devices will be manufactured and supplied? And, how are the processes and documents ensuring that this is done safely developed?
- How are unintended or nonconforming devices managed? And, how is quality monitored over time?

For any given type of device, the key documentation will be the risk assessment (described in section 7.3). Documentation may also include a general description of the device, device specifications (including limits such as maximum size), instructions for use, and standard operating procedures for manufacturing, packaging, storage, handling, distribution, monitoring of performance, cleaning and/or servicing, and disposal, where appropriate. ISO 13 485 [3] refers to this collection of reference documentation as a medical device file. The level of documentation used should be commensurate with the risk associated with the device (i.e. what hazard exists with errors in manufacturing or use) and the scale of device production (i.e. will the device be produced routinely). These documents are discussed in further detail in section 7.6.

Version control processes should be employed for these reference documents, with changes detailed and obsolete versions of documents archived for future reference. Documents should also be independently reviewed when drafted or revised, preferably by a multidisciplinary team of stakeholders, including the staff who would reference the documentation. For example, the instructions for use of a device must be reviewed by those who will be using the device.

It is good practice to record organisational experience in manufacturing devices for future reference, and for medical devices this will be required for regulatory compliance. This may include device requests, design iterations, manufacturing parameters, and results of quality control tests. Records should be kept not just of successful devices but also where failures have occurred, including printer or material failures, or issues with design, to inform future practice. Within a clinical radiation therapy centre, medical device records might be appropriately stored

Task
3D Device Structure Approved
3D STL Exported for Production
3D Print Preparation
3D Print and Post Processing Complete
3D Device Wax/Pb Production
3D Device Visual and Manual Inspection
3D Device Physics QA
3D Device labelled and Packaged
3D Device Delivered
3D Pre-Treatment Fitting Required
3D Device Code Capture

Figure 7.4. Example quality check list in an oncology information system.

within existing electronic medical record systems (e.g. attached to patient records in an oncology information system).

Figure 7.4 shows the process steps recorded as quality check-list items in the Mosaiq oncology information system of a radiation therapy centre producing patient-specific radiation therapy bolus in-house. In this example, electronic documents detailing completion of specific process steps (e.g. a device request signed by a medical professional) or the results of a quality control test (see chapter 8) would be attached to the patient's record.

7.5 Resource management

Resource management is a key component of a successful 3D-printing institution, service, or workflow. In the context of 3D printing, a resource is anything required to execute the accurate production of a 3D-printed device. Broadly, this can be split into two fundamental categories: human resources (Section 7.5.1) and infrastructure (section 7.5.2).

7.5.1 Human resources

Maintaining high-quality management of human resources requires organisation, delegation, and commitment to total quality management execution. Adequate staffing is required for any quality workflow in a radiation therapy centre. In the context of 3D printing, this will require a highly trained multidisciplinary team, including medical professionals, management and administration staff, and staff with expertise in manufacturing and quality assurance.

The number of staff required for a successful 3D-printing service will depend heavily on the scope of the service. Workforce recommendations [5] can be used as a task dependent guide for the number of staff required, taking an hour per activity estimation approach. This approach can be refined with thorough peer consultation with radiation therapy centres that already have 3D-printing services, who can provide guidance. Consideration should be given to the staff's skill sets and appropriate team numbers to ensure that the service is not weakened during staff leave. Moreover, one should remember that, as noted in chapter 6, the setup and commissioning phases of a 3D-printing service can be the most labour intensive. Once a self-sufficient system has been established, there can actually be a time saving component that may offset the requirement for additional staff [6].

Quality training is fundamental for empowered and confident staff in any context. Once documentation of workflows and use of equipment is completed (see section 7.4), this documentation can assist with staff training (e.g. a training guide may be created as a combination of workflow practices, vendor equipment instructions, and delegation of tasks). A thorough training programme will also include practical training, product and workflow familiarisation, and mock scenarios, as well as an option for feedback and review.

A clear delegation of tasks is required for smooth workflow and an efficient QMS. A multidisciplinary team meeting should provide space for all parties involved in the process of device creation to give input, and tasks should be delegated with this in mind. Total quality management (TQM) tools such as brainstorming, nominal group techniques (brainstorming with anonymity aspects), cause and effect diagrams, and responsibility assignment matrices may help with this. An example of a responsibility assignment matrix for a 3D-printing service creating bolus in a multidisciplinary team setting is shown in figure 7.5.

Visual representation of the workflow can be useful to avoid inefficiencies in multidisciplinary team task delegation and workflow creation (e.g. see figure 7.6). A visual flow chart of the workflow for a process can assist with avoiding bottlenecks, identifying where one team is carrying too much of the workload, and observing staff inefficiencies. Auditing this workflow and overall process management can help to identify unnecessary tasks, inefficiencies, and poor outcomes, all which can help guide management decisions.

7.5.2 Infrastructure

As noted in chapter 6, a successful 3D-printing service requires appropriate equipment and workspaces, to function safely and effectively. The management of infrastructure is an important part of the overall QMS. Consideration should be given to procurement method, storage management, sterilisation, data management, automation, resource management, appropriate space, and contingency planning.

Workspaces for manufacturing should maintain appropriate levels of cleanliness and hygiene for the production of devices intended for clinical use, as prescribed by relevant regulatory requirements. The 3D-printing process can produce free volatile organic compounds and ultrafine particles that should be addressed with appropriate filtration (e.g. on the printer housing itself) and with room ventilation.

Appropriate printer, filament, ancillary devices, and software should be determined with TQM at the forefront. To begin, stakeholders should outline an existing

Responsibility	Simulation RTs	Radiation oncologists	Planning RTs	Mould Room	Physics	Treatment RTs	3D printing experienced RTs
1. Requests	R/A	S	C	S	S	-	S
2. Planning	-	S	R/A	I	I	I	S/C
3. Fabrication	-	-	I	R	R/A	-	S
4. Wax bolus	-	-	R/A	R	S	I	S
5. Physics tests	-	-	I	I	R/A	-	-
6. Approval	-	I	R/A	-	C	-	S
7. Treatment	-	I	-	-	-	R/A	S
8. Reporting	I/C	I/C	I/C	I/C	I/C	I/C	R/A
9. Equipment	-	-	-	R	R/A	-	-
10. Training	C		-	-	R	-	R/A

Figure 7.5. RASCI matrix detailing staff (R)sponsible, (A)ccountable, (S)upporting, (C)onsulted and (I)nformed in an example 3D-printed bolus production workflow.

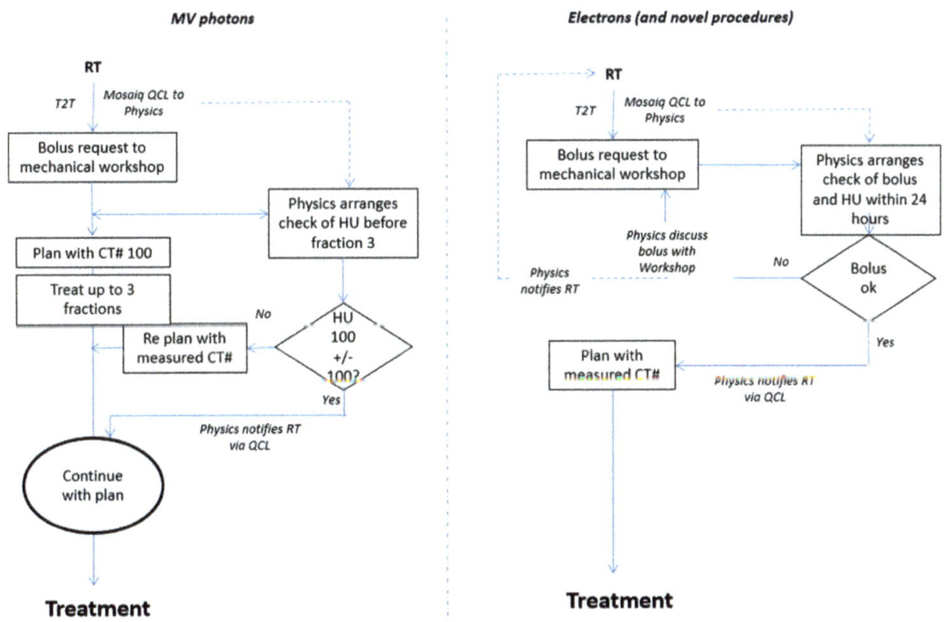

Figure 7.6. 3D-printed bolus workflow diagrams, where 'T2T' refers to an in-house work request platform called 'Track to Treat' and 'QCL' refers to a 'quality checklist' in the Mosaiq oncology information system.

need (which could be referred to as a purchase requisition) that is to be fulfilled by the purchase of goods. When determining appropriate equipment to purchase, a multi-disciplinary team approach is recommended. This is a key component of each team being able to perform their role adequately. Opportunities for staff to have input on appropriate equipment required to complete their delegated task should be given. Budgets and quotes can be created and ordered to assist with deciding on which equipment would be most suitable for purchase and which equipment would function to fulfil the existing need initially outlined. As per ISO 13 485 [3], records of results of all stages of procurement should be created and maintained, including evaluation, monitoring, and any necessary actions arising from these activities.

ISO 9001 [2] suggests that once the equipment has been purchased, processes are to be established to ensure safe practice, as exemplified in chapter 8. This can include recalibrations, ongoing quality assurance tests and maintenance, safeguards from adjustments that could affect the integrity of the equipment, and protection against damage or appropriate storage requirements. Appropriate tracking of stock levels for disposable items (such as filaments) will avoid supply issues. A minimum supply level should be determined initially and reviewed periodically to ensure that supplies are constantly stocked.

As discussed in chapter 6, appropriate storage should be considered for required stock and equipment. This also includes software or hardware storage. Any 3D printer should have appropriate bench space, storage for ancillary equipment and tools, and be in a location where access can be granted to required individuals with ease. Depending on the location of the 3D-printing equipment, appropriate ventilation will be required, such as a fume hood. As deemed appropriate, equipment safety should also be considered, in addition to providing restricted access to the room. Moreover, if sterilisation is required, then a robust sterilisation record-keeping process will be needed. ISO 13 485 [3] suggests that process parameters should be recorded for each batch and that records should be traceable.

Solutions for the recycling of unused or scrap filament should be built into the workflow when managing a new 3D-printing service (see chapter 6). The ability for a radiation therapy centre to recycle their own filament, or partner with institutes who do so, offers environmental benefits such as reduction in landfill and CO_2 emissions from waste transportation. Moreover, recycling filament can save costs in the long term. However, considerations in this space include the reduction of the molecular weight of the filament (such as polylactic acid or PLA) after excessive recycling and the need to add fresh (unused) filament to the recycling mix to mitigate this effect [7]. On a higher level, general environmental friendliness should be considered in the early stages of setting up a printing service, which may mean paperless record keeping where possible, minimisation of waste in the form of storage containers, and high-quality planning to reduce incorrect prints (see chapter 8).

7.6 Product realisation

Quality is achieved by developing processes that consistently produce results that meet the desired standard. The processes through which a device is produced (or product realised) should therefore be documented in a process map, device file, design or manufacturing plan, or case plan or summary. Generally, a process will include:

- Inputs, such as a specification of device requirements, a request from a clinician, regulatory requirements, or a risk assessment.
- Outputs, such as the resulting product and associated documentation.
- Verifications or validations, such as quality control tests that establish that the resulting product satisfies the initial requirements.

The product realisation process for a 3D-printed device can be broadly divided into six main stages: request, modelling/design, fabrication, post-processing and quality control. For more detailed discussion of some of these processes, see chapters 2, 4, 5, and 8.

7.6.1 Request and specification

The prevalence of 3D printing and the formidable rate at which it is becoming common practice in radiation medicine is attributed to its ability to provide an answer to a question, or fulfil a request, for a vast array of tasks. For a custom made medical device, this request must be made by a medical professional with documentation of this record. In some cases, this process may be iterative; for example, if the request cannot be 3D printed due to some limitation, there may be a revision of the request in line with what is achievable. This is where a multi-disciplinary team approach is useful.

A request for a device to be 3D printed may be made to fill a clinical need, serve as an ancillary device, act as quality assurance equipment, or assist with providing an innovative solution to a clinical question (see chapters 9 and 10). Due to this complexity and user preference, it is important to define the design requirements at the start of the product realisation process. Documentation on this should be detailed. For example, a radiation oncologist or therapist may request a bolus device with a specific relative electron density of 1 g cm^{-3}, smooth on the patient surface, of uniform thickness, and made of a flexible non-toxic material. These decisions should be recorded and kept, and any deviation from standard justified and recorded. Generally, these input parameters will be checked in quality assurance processes, and so clarity should be sought at this stage. This may include the rationale for the request, any specific features required (e.g. thickness, material, colour, etc), and a timeline for completion (see the example in figure 7.7). Setting up a process that requires this level of detail before moving on to the modelling or design stages will assist in achieving a consistently high-quality outcome.

7.6.2 Modelling and design

A detailed and considered design approach will allow the user to reap benefits later in the process of product realisation. A common theme in this space is the concept of 'Design for Manufacture and Assembly' (DfMA), which has traditionally been applied in the space of large scale construction and manufacturing jobs but can be

Parameter	Description
Type of model requested	☐ Bolus only OR ☐ Bolus and positive mould OR ☐ Positive mould only
Density	☐ 1.0 OR ☐ Other:
Materials	☐ PLA OR ☐ TPU/flexible OR ☐ Other:
Surface smoothing	☐ Yes OR ☐ No
Acceptance criteria for visual and manual inspection (T2)	Printed device matches device specification. No visible print defects (e.g. holes >3mm), no residual supporting structure. No noticeable loss of structural integrity. Suitable smoothness.
Acceptance criteria for dimensional inspection (T3)	@Define relevant definition for current model and manufacturing method, should ensure check of the critical reference structure dimensions e.g. Standard: Dimensional inspection via comparison of CT scan of device with approved model. Dimensions should be within image uncertainty (±0.5mm) for 90% of volume, and within ±2mm for all volume.
Acceptance criteria for radiological inspection (T4)	@ As specified by medical professional OR Standard: Radiological inspection via evaluation of CT scan Hounsfield numbers. Within image modality uncertainty and standard deviation in mean density; mean density should be within 0.1 of requested value, and standard deviation in mean should be < 0.2.
Device markings	@ Device ID, @any landmarks
Quantity for production	1
Delivery details	@ Location, @method of delivery (e.g. hand-over or post)
Electronic delivery details	@email
Further details	@NA

Figure 7.7. Example request for 3D-printed bolus.

applied to many workflows and applications. In essence, DfMA approaches may work to minimise the number of components in a design, ensure that tolerances are appropriate, and ensure that the design is clear and easy for a user to interact with. An example of this approach in a radiation therapy context is the design of a calibration jig, which may have multiple components that fit together. To ensure that the users are physically unable to set the jig up incorrectly, each attachment may have a differently shaped connector. In this way, the ISO9001 [2] stage of Design and Development Planning is achieved through DfMA processes.

Another concept in this space is lean planning and product development. Before launching into the development of a 3D product in a radiation therapy context, it can be useful to consider the target audience or final use of the product. From here, once can consider how the assembly line will operate and identify stages where waste can be reduced, or double handling can be removed. Lean production promotes team empowerment and multidisciplinary approaches, which are very relevant in the radiation therapy context.

The modelling and design stage should follow documented workflow processes that have been predefined, and remain in accordance with regulations and specified requirements. For example, the appropriate material suitable for different tasks should be identified, documented, and available (see chapter 3).

In most cases, where the printed product is deemed to be a medical device, explicit documented instructions should be made available. These instructions should have been made in consultation with a medical professional and should be free of ambiguity. It should not be possible to progress past this stage in the workflow unless documented approval has been given by more than one team member, adopting a defence in depth quality assurance approach. This approval should include the design explicitly being evaluated against the design requirements.

7.6.3 Fabrication

Once the detailed modelling and design plan is approved, fabrication can begin. While some of these models or plans may have been outlined in the modelling or design stage, this is not always the case. Variations in the environment and printer itself can contribute to uncertainty and changes in the output of the fabrication stage, and this should be monitored and acted upon during production. A key consideration here is process monitoring because it may not be economical to wait until the end of a print to ensure success. This may mean setting up a periodic checking roster for prints (particularly when they are to be clinically rolled out). This is particularly relevant for large prints. In the context of radiation therapy, this may include prints of a patient's face for a shield design or large bolus prints (see chapter 9).

In some cases, remote monitoring of print progress may be possible, depending on the type of printer and its ability to send video footage over a local network. As always, care should be taken to ensure that this virtual monitoring cannot reveal patient identifying information or treatment data, including information shown in the name of the print job or embossed into the print surface, outside the radiation therapy centre for which the print is produced.

If an error occurs during the print, then it is beneficial to identify this as early as possible and assess the effect of the error. In the fabrication stage, it can be highly beneficial to ensure that detailed documents are available for users and that these should be available as online copies but also on hand near the printers to assist with troubleshooting. Appropriate documentation of failures or non-conformant devices that do not meet design specifications should be kept because they can be a key tool to learn from mistakes and errors, and adjust the process accordingly (see chapter 8).

7.6.4 Post-processing

In 3D printing, the equipment that is used to achieve desired results can vary greatly (see chapters 2 and 3). Nevertheless, most combinations of printer, filament, and production line will still require a basic post-production step (see chapter 5), be it removal of the support structure or brim layer. Post-processing can improve the function, aesthetic, or characteristic of the print. This may include sanding, burring, cleaning, painting, polishing, and potentially coating. Depending on the function of the design, post-processing may be a key step to the quality of the print.

A key example of this is for bolus that may be used on exposed or sensitive patient surfaces. If the print is not smooth on the patient's side, then it will be uncomfortable for the patient and an overall low quality design. However, if this post-processing step is included, then the overall quality of the print is significantly increased.

This stage of the product realisation will also require a stringent labelling system to be in place, which is particularly important for clinically-based prints for patient use. Regulatory bodies may provide guidance on the type of labelling required for a medical device (GMDN, see section 7.3) and local protocols can be added to this (i.e. labelling patient-specific devices with patient, treatment, and anatomical site identifying information).

7.6.5 Quality assurance

The specifics of quality assurance for 3D printing in a radiation therapy context will be addressed in detail in chapter 8. This section describes the place of quality assurance in the broader QMS.

On a higher level, design and development verification (as outlined in ISO 9001 documentation [2]) is the process of assessing the design output and ensuring that all of the design inputs have been considered. For the radiation therapy bolus example that is shown in figure 7.2, the geometry of the bolus will be optimised for the patient in a treatment planning system using medical image data (input), the structure will be approved by a medical professional (validation), an STL file will be produced for 3D printing with a suitable material (input), resulting in a patient-specific bolus (output) for which geometric and density checks will be performed (validation). Each of these steps will involve their own processes and documentation.

Quality control testing should form part of the quality assurance system. While quality control and quality assurance are frequently used synonymously, maintaining the distinction defined by the ISO [1, 2] is useful when setting up a 3D-printing service. Quality assurance is process-focused and aims to *avoid* defects, whereas quality control is output-focused and aims to *detect* defects. As described in chapter 8, quality assurance encompasses the commissioning and initial setup of the quality control programme, while quality control involves the routine testing of specific outputs, such as 3D-printed treatment devices or test objects. Outcomes and trends can be monitored by tracking the quality control test results.

The identification and consistent use of appropriate tolerances for quality control is an important part of quality assurance and should be included in the quality assurance workflow. Ideally, initial tolerances would be borrowed from published guidelines or codes of practice. However, in the absence of widely-accepted recommendations (while, for example, recommendations from AAPM Task Group 336 are currently in preparation), each 3D-printing service must identify its own initial tolerances. This can be achieved by evaluating the treatment effects of common 3D-printing errors and using that information to determine clinically acceptable limits [8] (see chapter 8). After a period of tracking quality control results, it may be possible to tighten local tolerances using statistical process control methods or by quantifying uncertainties in the quality control process. These uncertainties might be defined by device resolution or anatomical image resolution, for example. This can help to align the product outputs to the product requirements, but more importantly it can assist radiation therapy centres in ensuring that their 3D-printed devices will be consistently reliable, accurate, and safe for use. Action levels will be based on the ranges that are acceptable for use, outside which the device or product cannot be used. This might include a Hounsfield unit range for clinically acceptable dosimetry, for example.

Overall, quality assurance helps to ensure that each step of the product realisation is high quality. An example of a quality assurance tool that covers the complete process of product realisation is the Failure Modes and Effects Analysis (FMEA) or System Theoretic Accident Model and Processes (STAMP). A FMEA is created by identifying all of the ways in which something could go wrong in a process to identify weaknesses. The risk is assessed with each potential failure and succinctly depicted in an FMEA analysis. A FMEA can be created from a process map and the user can expand on all possibilities of erroneous efforts or incorrect actions leading to failures. Further advice on creating an FMEA can be found in various guidelines, such as the AAPM Task Group 100 report [4].

7.7 Ongoing responsibilities

Reviewing workflows, learning from mistakes, and contemplation on the efficiency and efficacy of 3D-printing processes are as important as any quality assurance measure. Recording and monitoring outcomes, also referred to as post-market surveillance, refers to the act of reviewing outcomes and revising work practices accordingly. This is an aspect that should not be overlooked when initially implementing a 3D-printing service. To constantly improve in this space and execute a high-quality 3D-printing service requires constant revision, communication, and asking 'how can this practice be improved?'

One example of recording and monitoring outcomes would be looking into quality control data and performing statistical process control. Based on the findings from such a process control tool, an appropriate action can be taken to further improve and enhance the effective and efficient 3D-printing service.

Feedback is vital when initially implementing a 3D-printing service. Feedback from stakeholders, including multidisciplinary teams and patients, is an abundant resource that should be utilised. Feedback itself is a great measure of effectiveness [3] and should be sought during production and post-production. Any robust feedback programme will include a process for handling complaints, which is also outlined in ISO 13 485 [3]. All complaints should be recorded and collated to identify common or recurrent problems, and acted upon to improve the practice. Knowledge of local regulatory authorities and reporting requirements is necessary to ensure safe practice. It is also vital to report all non-conformance and take corrective action with an appropriate authority, whether that is a local safety committee or an external body.

Documentation developed for each step in the product realisation chain should be periodically reviewed for adequacy and accuracy to ensure that it is reflective of current practice. Ongoing reviews should use the evaluation to inform necessary changes or action items, which should be clearly communicated to responsible team members (see section 7.5.1).

To optimise and ensure a high-quality 3D-printing service, regular training should be scheduled and conducted to maintain a high skill level among staff. Instances of accidents and near misses have been reported as a result of lack of training in the radiation protection literature [9]. It is important to ensure that training is provided by competent staff who are themselves highly trained to minimise the potential for accidents. If relevant, incorporating such training into orientation packages can be an option because the training schedule may not align with start dates for new employees. Regular training sessions are an excellent opportunity to educate users on any improvements or changes that were made to the overall 3D-printing service.

For a 3D-printing service in a radiation therapy centre to be effective, it is important to establish regular preventative maintenance inspections, and servicing of the printer and associated equipment on a regular frequency (see chapter 8). Work practices and procedures should also be reviewed regularly and updated as required. Obviously, 3D-printing quality assurance processes and outcomes should also be reviewed regularly, with a view to achieving continuous quality improvement. Ideally these reviews and recommendations should be part of the quality management committee's responsibilities and must be communicated to appropriate user groups. Any regulatory reporting requirements should also be maintained, which is the responsibility of the quality management committee. The multidisciplinary team approach to 3D-printing quality management should be retained throughout the lifetime of the service.

7.8 Key points

- All devices printed in or for a radiation therapy centre should be created, assessed, documented and used in compliance with all applicable national or international regulations or standards, which often require the creation of a QMS
- A formal QMS is needed for medical devices, and is advisable as good practice for all 3D printed objects created in a radiation therapy centre
- The QMS defines processes, documentation and staff responsibilities and is a valuable tool for achieving continuous quality improvement in the 3D printing service
- The QMS should be created and maintained by a multidisciplinary team representing all professional groups in the radiation therapy centre and should include a statement of the centre's commitment to quality
- Adherence to the QMS should ensure that 3D printed devices perform as intended, are safe to use, and have benefits that outweigh their risks, while also providing a framework for establishing adequate staffing levels, providing staff training and determining the methods, tolerances and frequencies of routine quality control, audits and reviews of the service and its output.

References

[1] International Organization for Standardization 2015 *ISO 9000—Quality Management Systems—Fundamentals and Vocabulary* [Online] Available: https://www.iso.org/standard/45481.html (visited on 06/01/2023)

[2] International Organization for Standardization 2015 *ISO 9001—Quality Management Systems—Requirements* [Online] Available: https://www.iso.org/standard/62085.html (visited on 06/01/2023)

[3] International Organization for Standardization 2016 *ISO 13 485—Medical Devices—Quality Management Systems—Requirements for Regulatory Purposes* [Online] Available: https://www.iso.org/standard/59752.html (visited on 06/01/2023)

[4] Huq M S *et al* 2016 The report of Task Group 100 of the AAPM: application of risk analysis methods to radiation therapy quality management *Med. Phys.* https://doi.org/10.1118/1.4947547

[5] International Atomic Energy Agency 2015 *Staffing in Radiotherapy: An Activity Based Approach* Vienna [Online] Available: https://www.iaea.org/publications/10800/staffing-in-radiotherapy-an-activity-based-approach (visited on 06/01/2023)

[6] Albantow C E and Brown S J 2022 Three-dimensional printer use in the Australian and New Zealand radiation therapy setting *J. Med. Radiat. Sci.* **69** 492–501

[7] Zhao P, Rao C, Gu F, Sharmin N and Fu J 2018 Close-looped recycling of polylactic acid used in 3D printing: an experimental investigation and life cycle assessment *J. Cleaner Prod.* **197** 1046–55

[8] Kairn T *et al* 2021 Determining tolerance levels for quality assurance of 3D printed bolus for modulated arc radiotherapy of the nose *Phys. Eng. Sci. Med.* **44** 1187–99

[9] International Atomic Energy Agency 2000 *Lessons Learned from Accidental Exposures in Radiotherapy* Vienna [Online] Available: https://www.iaea.org/publications/5818/lessons-learned-from-accidental-exposures-in-radiotherapy (visited on 06/01/2023)

Chapter 8

Quality assurance

Adam Unjin Yeo and Tanya Kairn

8.1 Introduction

This chapter focuses on the quality assurance processes that are involved in setting up and maintaining a service that is capable of reliably creating clinically acceptable three-dimensional (3D) prints for radiation therapy applications. As noted in chapter 7, quality assurance is a broad area that is focussed on *preventing* defects, whereas quality control tests specific outputs to *detect and identify* defects. Consequently, this chapter on quality assurance includes descriptions of common 3D-printing defects and their potential effects on the treatment accuracy of radiation therapy, explanations of the commissioning steps aimed at systematically preventing defects, and examples of quality control methods for detecting defects in specific 3D-printed radiation therapy devices.

As discussed further in chapters 9 and 10, a wide range of radiation therapy treatment equipment can be beneficially augmented or replaced through the use of 3D-printing. This equipment includes boluses and compensators [1, 2], radiation shields [3–6], immobilisation equipment [7–9], and brachytherapy applicators and templates [10–13], as well as dosimetry jigs [14], accessories [15], and measurement phantoms [16–18].

When these beam-modifying, patient-immobilising, or dose-measurement devices are fabricated using an in-house 3D-printing service (or even using an external 3D-printing supplier [19]), the accurate construction of these devices with respect to material homogeneity, physical density and geometry must be demonstrated and documented via dedicated quality assurance processes.

It should be understood that common 3D-printed objects such as ornaments, toys, jewellery, architectural mock-ups, mechanical part designs, and even surgical planning models usually involve the production of low-density prints with aesthetically-pleasing surfaces. By contrast, 3D-printed accessories and patient-matched devices used in radiation therapy centres must achieve different standards. Radiation therapy 3D prints are usually:

doi:10.1088/978-0-7503-3907-0ch8

1. Printed at comparatively-high infill densities (e.g. shielding, bolus, applicators, phantoms); and/or
2. Expected to achieve consistent geometric accuracy across long, thin, curved structures (e.g. shielding, bolus); and/or
3. Very smooth on all surfaces, including surfaces abutting scaffolding (e.g. bolus, immobilisation devices, jigs, applicators); and/or
4. Designed to be comparatively large, printed in parts that must fit together, or have thin or tapering edges (e.g. phantoms, shielding, phantom components, applicators); and/or
5. Able to fit together extremely well or otherwise achieve strict geometric accuracy (e.g. jigs, attachments, adaptors, phantom components); and
6. Manufactured with excellent integrity without unintended deformity (due to heat, moisture, radiation damage, etc), non-uniformity, holes, or unintended air gaps.

Due to the additional (and often stricter) requirements of 3D printing for radiation therapy applications, carefully planned and documented quality assurance processes are needed in this area. This chapter starts by building on the fundamental information provided in chapter 5 by summarising the defects that commonly arise in 3D printing for radiation therapy, how each defect could be minimised, and how the defect would otherwise affect radiation therapy accuracy (section 8.2). This chapter then provides guidance for the development of quality assurance processes for avoiding, minimising, detecting and identifying those defects, through the use of commissioning (Section 8.3), 3D-printer maintenance (Section 8.4), and 3D-print quality control (section 8.5).

Additional testing needed when commissioning 3D-printing devices for use in patient treatments is discussed in chapter 9, and further testing of 3D-printing devices to verify treatment plans and treatment delivery systems is discussed in chapter 10.

8.2 Defects and consequences

8.2.1 Irregular surface

Unintended irregular surfaces (see figure 8.1(a)) may be caused by many factors, such as unsmoothed steps in the segmented structure (due to computed tomography (CT) slices) from which the 3D-print design was derived (see chapter 4), residual scaffold after inadequate post-processing (see chapter 5), or 3D-printing issues such as stringing, excessive layer height for filament thickness, and print speed (see chapter 5). These issues can also manifest as a collapse or poor quality of an overhanging surface, which leaves imperfections such as small bulges or beads.

Common problems during the printing process are that the filament is too hot when extruded or the filament cooling system has not been efficient enough. This can also cause issues at overhangs where the over-heated filament does not solidify quickly enough to avoid being dragged into an unwanted location before solidification. This phenomenon repeats or increases from one layer to another. Potential

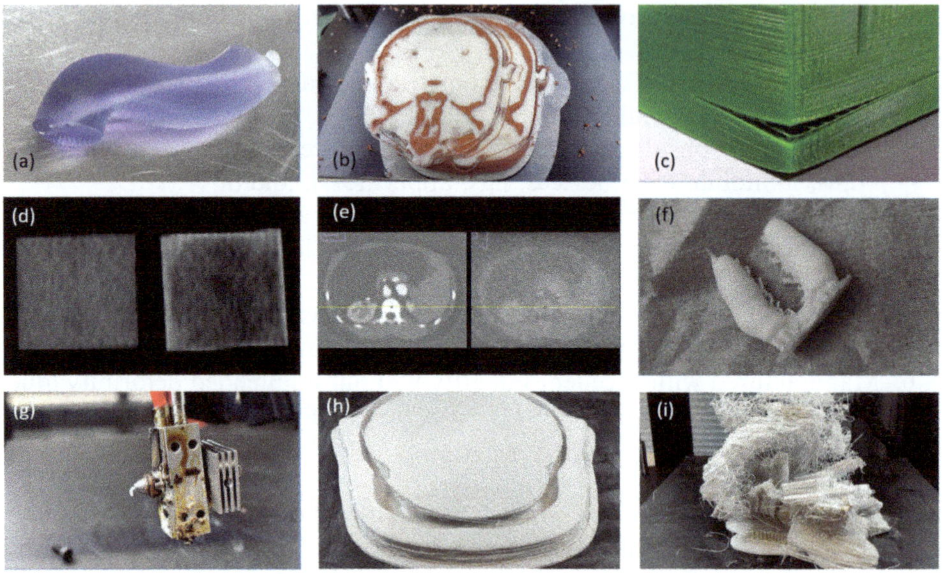

Figure 8.1. Examples of common issues affecting 3D prints for radiation therapy purposes: (a) unintended surface roughness on a vat photopolymerisation print of a brachytherapy applicator after removal of support structures; (b) offset discontinuities resulting from multiple lateral shifts of a material extrusion (MEX) print head during fabrication of a slab phantom; (c) deformation of a printed slab due to thermal warping and detachment from print-build plate; (d) CT slices showing a 3D print with consistently correct density (left-hand side) and with internal density heterogeneity (right-hand side); (e) CT slices showing material doping that realistically represents tissue and bone at kV energies (kV CT left) but not at MV energies (MV CT right); (f) filament stringing in the empty space between two sides of a nasal packing print; (g) disassembled MEX print head showing damage from an extended pause in printing; (h) slice of a head phantom printed with a gap where the skull should be after the bone-equivalent filament snapped during printing; (i) birds nest arising from MEX printer continuing to print after an object had become detached from the print-build plate.

preventative methods are to reduce the nozzle's temperature, change the printing orientation to minimise overhangs (see chapter 5), or to upgrade the 3D printer with a better cooling system.

Rough surfaces can also be caused by overly thin shell layers (sometimes called 'walls') on the surface of 3D-printed objects. This could result from contours being too thin or not fully covered or enclosed. Not depositing enough material could also cause an irregular surface, resulting in a shifted layer on the surface or sharp edges at the corners. In such cases, the 3D-printing parameters (especially temperature settings) and the 3D-printer's performance should be checked. For example, for material extrusion (MEX) printers (aka FDM or FFF printers, see chapter 2) the filament extrusion rate may need to be recalibrated so that the filament is flowing appropriately at the shells (walls) and throughout the print.

If not corrected, unintended irregular and rough surfaces are issues for any printed objects, but can cause problems for radiation therapy patient treatments (e.g. discomfort, pain, exacerbated skin reaction, etc), especially if the device needs to be in contact with sensitive patient surfaces (e.g. mouth, lips, buccal mucosa, especially after a couple of weeks of head-and-neck radiation therapy).

8.2.2 Geometric error

Geometric errors in 3D-printed devices have the potential to cause systematic errors in treatment dose delivery if poorly fitting devices are used on patients, leaving unintended air gaps or altering the path-length of the radiation beam. This uncertainty can also unintentionally increase susceptibility to treatment setup variation.

A common cause of a geometric error being introduced during 3D printing is that an inadvertent offset of a layer along width or length axes (or both) is introduced by moving the print head or platform during the printing process (see figure 8.1(b)). This may be corrected by reducing the acceleration on the axis affected by the problem.

Another well-known cause of geometric uncertainties is thermal warping or curling of corners, typically along the height direction (see figure 8.1(c)). This effect can result from temperature gradients within the printed object (which is more common for larger and denser prints), differential shrinkage of regions of the printed material (i.e. excessive material shrinkage coefficient), and poor adhesion between the base of the object being printed and the build plate (sometimes called the 'print bed') to which it should be attached, which allows the print to flex in the vertical direction.

Potential correction methods for MEX printing are:

- To verify the height and levelling of the build plate before the start of printing (and adjust if necessary);
- To alter printing orientation in order to increase contact between the base of the print and the build plate, and reduce overhang;
- To change the printing material (e.g. warping occurs less with polylactic acid (PLA) than other softer materials);
- To increase the build plate temperature in conjunction with adding adhesive material to the print-build plate (or instead cleaning the print-build plate for some systems); and
- Most importantly, to change the infill pattern (see chapter 5), at least for the first layer of the print (e.g. filling the first layer with a concentric rather than linear pattern and then filling the inside with honeycomb can prevent shrinkage).

8.2.3 Density variation

Density variations that can occur in 3D-printed objects (see figure 8.1(d)) include the mass density or relative electron density (RED) being consistently greater or less than intended, the mass density or RED varying throughout the print, and voids or other defects occurring within the print.

A common cause of density variations is non-homogeneous deposition of material, with interdependent contributing factors such as inconsistent material flow rate, temporary absence of material, inconsistent nozzle temperature, and varying cooling speed during printing process. Density heterogeneities can also be emphasised by designing objects that are very thin, so that the 2–3 mm shell thickness (which is usually denser than the internal material, even when printed at 100% infill [17] makes a comparatively large contribution to the overall density assessment.

Potential correction methods for MEX systems are:

- To print test samples, and reassess the purity and integrity of the material itself (which can vary between samples [20]; even a trusted supplier can produce a 'bad batch' of filament or a composition can be deliberately varied for production or supply chain reasons);
- To examine the filament drive system and see if the filament is blocked upstream of the extruder;
- To clean the nozzle to enable consistent flow of material or replace it with a new nozzle;
- To check the temperature setting and the temperature control system; and
- To increase the thickness of the print design, because it is often difficult to print a thin object with homogeneous density.

Unexpected changes in print density can also occur due to upgrades in the 3D-printer's firmware [21]. Consequently, frequent quality control checks of the 3D-printing process (via periodic reprinting of test objects, see section 8.5) is advisable.

Unobserved and uncorrected density variations can affect most radiation therapy applications, although they are of greatest concern when the 3D-printed device has the potential to affect the radiation beam (e.g. beam-modifying devices) and may be less concerning when the device is a jig, accessory or support, where dosimetric effects are not expected.

8.2.4 Unsuitable doping

Unsuitable doping is a broad term that was coined to refer to situations where a device is not suitable for use in the radiation beam for which it was intended or where a device that was created for use at one beam energy is used at a beam energy for which it is not suited.

Due to the widespread availability of kilovoltage (kV) CT imaging in radiation therapy centres, unsuitable doping in the 3D-printing context usually occurs when a doped composite material is found to have a tissue-like appearance at the kV imaging energy but lacks the corresponding tissue-like effect on the megavoltage (MV) treatment beam.

For example, Crowe *et al* [22] investigated the unusual case of a scalp radiation therapy patient who had a bone flap replaced with a 'bone cement' made from plastic doped with barium sulphate. The barium sulphate increased the effective atomic number of the cement but had a minimal effect on the material's RED, causing the cement to look like bone in the treatment planning CT image and yet look like plastic in the MV CT images acquired at treatment, and also causing plastic-like (not bone-like) attenuation of the MV treatment beam [22].

Although a wide range of doped composite materials are useful for radiology education and research due to their bone-like appearance in kV images [23], the different effects of each one of these materials in MV therapy beams would need to be verified prior to use for radiation therapy purposes. Figure 8.1(e) shows kV and MV CT scans of a phantom made using a vat photopolymerisation (VPP) process from flexible resins doped to look like different human tissues in kV CT images, where these materials do not have corresponding effects on an MV imaging (or treatment) beam. Notably, the materials used to 3D print the example shown in figure 8.1(e), RadioMatrix (Stratasys, Rehovot, Israel), promote these doped resins as suitable for medical imaging and not radiation therapy.

Clearly, this issue is not a problem if phantoms designed for their radiological properties in kV imaging beams are used only in kV imaging beams. The doping only becomes 'unsuitable' when the phantoms are used in MV therapy beams. Consequently, the problem of unsuitable doping can be avoided through communication and administrative controls, so that phantoms are not used inappropriately.

If phantoms are used inappropriately, then there may be serious downstream effects on the radiation treatment accuracy. For example, if a phantom containing unsuitably doped 'bones' is used to evaluate MV treatment accuracy, then dose differences in and around the bones may be attributed to the treatment planning system and not to the phantom (which will look entirely suitable when examined with a kV CT scanner). In the worst case, changes to treatment planning data may be made (or not made) based on the results of measurements that are unreliable. This has the potential to affect all patients subsequently treated using that treatment planning system (TPS).

At the phantom design stage, unsuitable doping of MV treatment phantoms can be avoided by evaluating all potential phantom materials using MV imaging or MV transmission measurements [17, 24], so that each material's validity for the intended purpose can be established before the phantom is 3D printed.

Work by Hill *et al* [25] has indicated the potential for this problem to occur in the opposite direction—water-equivalence (or tissue-equivalence) of materials in kV therapy (or imaging) beams should not be assumed based on the same material's demonstrated water-equivalence (or tissue-equivalence) in MV therapy beams.

8.2.5 Bulk deformation and clearance

Although material deformations due to temperature gradients are covered in section 8.2.2, materials can also deform more uniformly as they cool after printing. This bulk expansion or contraction is a common cause of 3D-printed parts failing to fit into their intended positions in existing equipment or failing to interlock with other 3D-printed parts correctly. Materials should be evaluated at the commissioning stage (section 8.3) to identify systematic expansions or contractions and identify relevant margins to use when designing interlocking parts.

Another cause of clearance issues is a lack of precision from the 3D-printer itself. Clearly some systems (e.g. VPP) provide better precision than others (e.g. binder jetting and some types of powder bed fusion), as described in chapter 2. However, even for MEX printers, differences in precision can be observed between different printer models as well as different printing settings (see in chapter 5). As with materials, printing of test objects at the commissioning stage can be used to detect issues with 3D-printer precision and identify relevant margins for interlocking parts.

For MEX printers, an issue affecting clearance can arise from the 3D-printing process itself, where thin threads of filament are woven in the empty spaces between the different parts of a 3D-printed object. This effect, which is typically called stringing or oozing (see figure 8.1(f)), occurs as plastic continues to flow from the print head as it moves, due to residual pressure in the heating element and the fluidity of the molten plastic.

It is possible to physically remove unwanted strands of plastic after printing, but obviously it is preferable (e.g. less wasteful of time and material) to minimise their production. Several steps can be taken to minimise the production of stringing or oozing, which should be investigated at the commissioning stage:

- Increase the withdrawal length of the filament (called 'retraction length' or 'retraction distance', depending on the slicing software). The recoil of the filament will cause the pressure in the print head heater to drop. This effect can be modulated by also altering the withdrawal speed.
- Increase the print head travel speed. This leaves less time for the molten plastic to flow and leave traces between the printed parts. Note that printing speed affects accuracy (see chapter 5), which might be counterproductive when attempting to improve clearance, so this parameter should be optimised.
- Reduce the extrusion temperature of the plastic. If it is too high, then the plastic is more fluid, and therefore escapes more quickly through the extruder. Given the effects of temperature on other forms of print deformation (section 8.2.2), this parameter should also be optimised to achieve best results.

Issues with integrity and clearance can affect jigs, attachments, adaptors, and parts that are intended to fit into existing equipment (see the examples in chapter 10), which require precise fabrication of interlocking parts. The clinical effects of bulk deformation and clearance defects largely relate to wastage of material and unproductive use of staff time.

8.2.6 Print failure

A print failure occurs when a 3D-printing job is started but fails to run to completion. A total failure of a 3D-printing job does not have quality control implications because the failed print will simply replaced rather than used. However, print failures should be recorded and tracked due to quality assurance implications. If similar failures recur, then changes may need to be made to the printing process or additional staff training may be required.

Print failures can occur when the printer loses its connection with the printing material. For MEX printers, the usual reasons for this are that the filament snapped during printing, tangles occured on the spool, or the filament supply ran out before print completion. This type of failure may be either recoverable or unrecoverable, depending on the timeframe and type of printer involved.

Some MEX printers can detect when the filament stops reaching the print head and will pause printing until the problem can be resolved. However, this may lead to further issues, including damage to print head components if the hot end is kept hot during a long pause (e.g. overnight or over a weekend, see figure 8.1(g)). The loss or lack of a filament may also be recovered on less advanced MEX printers if they are closely attended because a new filament can be pushed into the hot end as the previous filament runs out.

Print failure from loss of filament can become unrecoverable if the 3D printer does not detect the loss of the filament and continues 'printing' as though the filament is still present. This can lead to surprising outcomes during dual (or multi-) filament printing, if one filament is lost while the other(s) continue printing, as exemplified in figure 8.1(h).

Another very common and visually remarkable indicator of 3D-print failure is the production of a 'bird's nest', as shown in figure 8.1(h) (also see chapter 5). Generally, this occurs during MEX printing when the base of the object being printed becomes detached from the print-build plate and begins to move around, while the printer continues depositing filament as though nothing was wrong. If not attended, the print job will run to completion but the result will be clearly unsatisfactory.

Methods for improving print adhesion to the print-build plate should be identified during commissioning because remedies are likely to depend on printing material, temperature, and type of build plate surface. In some cases, covering the print-build plate with a layer of water-based glue can improve adhesion, though in other cases the best remedy is to carefully clean the build plate. Other strategies that may be employed include surrounding the first layer of the print with a wide brim of additional printed material, adding additional support structures, or changing the orientation of the object so that better contact with the build plate is achieved (see chapter 5).

A loss of adhesion (and consequent bird nest) may also occur as a result of inadequate print-build plate levelling, which causes one side of the print to have reduced contact with the build plate, or print warping (see figure 8.1(i), also see section 8.2.2) can cause one side of the print to lift up. The side of the print with poor contact, or the side that has risen up due to warping, may then obstruct the path of the nozzle, causing a collision and additional pressure on the print each time that the nozzle passes the elevated region, ultimately leading to total detachment.

While extensive test printing can help to identify the print settings that minimise the likelihood of failure, the key commissioning step that is most likely to reduce these types of print failure (i.e. insufficient, tangled, snapped filament, and loss of adhesion to the print-build plate) is the provision of detailed instructions via documentation and staff training.

8.2.7 Consistency and reproducibility

Limitations on consistency and reproducibility of 3D-printed radiation therapy devices can arise from unexpected changes in the 3D-printer's performance (e.g. firmware upgrades or faulty parts), changes in the 3D printer's use (e.g. printing with untested filaments, printing unusual geometries, or deviating from the printing parameters recommended via commissioning), as well as variations in printing material (e.g. from exposure to water, humidity, or very high or low ambient temperatures, or from unreported changes in composition). All of these changes have the potential to affect the accuracy of radiation therapy treatments that rely on 3D-printed devices.

Consequently, it is important to routinely evaluate the 3D-printer's performance via routine maintenance checks and quality control testing, evaluate material consistency via commissioning tests and quality control testing, and communicate approved 3D-printing methods via staff training and written instructions. These quality assurance tasks must be in place for routine use of 3D printing in radiation therapy centres (see sections 8.3, 8.4, and 8.5).

8.3 Commissioning

8.3.1 Risk assessment and overview

Having summarised the frequent 3D-printing defects and their potential effects in the radiation therapy context in section 8.2, this section will broadly describe the steps involved in planning, completing, and documenting a commissioning process for a radiation therapy 3D-printing service. Here, it is assumed that other aspects of quality management (e.g. completing a risk assessment as well as planning, resourcing, providing space and staffing the service, see chapters 6 and 7) have already been completed, in addition to procuring the 3D-printing equipment and performing customer acceptance testing. The 3D-printing workflow (from request to clinical use) should also have been finalised by the multidisciplinary team responsible for 3D-printing quality management, as described in chapter 7.

Nonetheless, a very brief and nonspecific risk assessment summary is included in table 8.1, based on defects described in the previous section, to provide an example and a basis for the selection of commissioning steps in this section. Obviously, a comprehensive local risk assessment should be completed for each individual 3D-printing service to take local equipment and local treatment needs into account.

If table 8.1 was an accurate summary of the risk assessment completed for a hypothetical new radiation therapy 3D-printing service, then the steps listed under the 'mitigation' column, as well as documentation and training that are listed in the caption, would describe the work required to reduce all risk levels for the hypothetical service to either 'low' or 'negligible'. Based on those mitigation tasks, the key commissioning requirements arising from the risk assessment in table 8.1 would be as follows:

- Optimise 3D-printing parameters;
- Evaluate geometric accuracy;
- Characterise physical properties of 3D-printing materials;
- Characterise density and Hounsfield unit (HU) properties of 3D-printing materials;
- Evaluate 3D-printing consistency and reproducibility;
- Plan a schedule of routine maintenance for 3D-printing hardware;

Table 8.1. Example risk assessment summary for a radiation therapy 3D-printing service. Documentation and training have been omitted from the mitigation column because they apply to every cause. *Not for uncritical use—all data in this table are hypothetical and arbitrary. Every 3D-printing service must complete their own risk assessment based on local circumstances.*

Cause	Hazard	Frequency	Severity	Risk level	Mitigation
Surface irregularity	Patient discomfort/pain, bacterial growth	High	Med	High × Med	Parameter optimisation, post-processing, sanitisation procedure
Geometric error	Patient discomfort/pain, setup error, air gaps, dose perturbation	Med	High	Med × High	Parameter optimisation, printer maintenance, quality control
Density error	Dose perturbation	High	High	High × High	Parameter optimisation, material commissioning, quality control
Density non-uniformity	Dose perturbation, hot/cold spots	Med	High	Med × High	Parameter optimisation, printer maintenance, quality control
Void (internal air gap)	Dose perturbation, hotspot	Low	High	Low × High	Parameter optimisation, printer maintenance, quality control
Unsuitable doping	Dose-measurement error	Med	Med	Med × Med	Material commissioning, administrative controls
Unsuitable doping	TPS adjustment error	V.Low	V.High	V.Low × V. High	Material commissioning, administrative controls
Clearance fault	Bulk deformation or surface irregularity	Med	Low	Med × Low	Parameter optimisation, Printer testing, material commissioning
Print process inconsistency	Systematic errors in treatment delivery	High	High	High × High	Comprehensive commissioning, maintenance, quality control, documentation

- Develop and document 3D-print sanitisation processes;
- Develop and document 3D-printer and 3D-print quality control processes;
- Document commissioning outcomes, including all recommended printing equipment, materials, and parameters, as well as all identified 3D-printing limitations and restrictions; and
- Develop and deliver a programme of staff training regarding 3D-printing safety, 3D-printer maintenance, 3D-print preparation, recommended printing parameters, 3D-print sanitisation, and fault reporting.

It is important to also add to that list the completion of a preliminary literature review (including reading all instructions provided with 3D-printing hardware and materials); test printing of challenging geometries, including standard test objects [26–28]; test printing of long print jobs if there is an intention to leave the 3D printer (s) running out of hours or overnight; completion of end-to-end testing of the entire process of requesting, printing, checking, and using 3D prints; setup of logs of 3D-printing activity, maintenance, sterilisation, and faults; and a requirement to revisit all of these steps whenever new 3D-printing equipment or processes are introduced into the service. Each of these commissioning steps is briefly described below.

8.3.2 Familiarisation with existing documentation

As with any other commissioning process, it is advisable to start by reviewing all of the available documentation on the equipment and materials that are to be used in the 3D-printing service. This should include manufacturer documentation (e.g. user guides, technical specifications, and safety hazards), as well as scientific literature regarding the specific use of this equipment for the intended radiation therapy purpose. One should not overlook the importance of this step as a starting point for gaining insight into the characteristics and limitations of the given system(s).

8.3.3 Optimisation of the 3D-printing parameters

This step may be completed iteratively with subsequent steps, as materials and potential print geometries are investigated. However, this step needs to be completed systematically, with studious recording of results, to minimise wastage of materials. (Inevitably some wastage is unavoidable because many prints should fail during this process.)

Broadly, this step evaluates the effects of varying all accessible 3D-printing parameters, for each combination of 3D printer and material, and thereby identifies an optimal selection of parameters that reliably achieve the properties required of the 3D-printed devices. Optimisation goals will depend on local conditions and priorities. For example, for 3D-printed bolus, speed, thickness, and density consistency may be prioritised above fine-scale precision, whereas fine-scale precision is likely to be prioritised over print density for the connecting parts of a measurement jig.

The user-editable printing parameters for MEX that were identified (and described in detail) in chapter 5 are nozzle temperature, build plate temperature, printing speed, layer height, infill density, infill pattern, build orientation, and design of supports, but access to these parameters may depend on slicing software, and different sets of user-editable parameters may apply to the various other types of 3D-printers described in chapter 2. These parameters should be adjusted, investigated and optimised with reference to local 3D-printing goals, as well as the common 3D-printing defects described in section 8.2.

Documentation of final recommended parameter sets (for each 3D printer and printing material combination), as well as details regarding how those parameters were identified, will play a critical role in any procedure reviews that are conducted under the quality management system (see chapter 7), while also providing a basis for guiding and streamlining the commissioning of future 3D printer and printing material combinations.

8.3.4 Evaluation of geometric accuracy

To evaluate geometric accuracy, including clearance, a range of test objects should be printed and their dimensions should be measured with callipers or a purpose-designed clearance gauge. The dimensions of these test objects should extend below the minimum and above the maximum thickness expected in the clinical 3D-printing service. Geometric accuracy should also be assessed across the range of materials and densities that are expected in the clinical 3D-printing service.

For future design of jigs or interlocking parts, this commissioning step provides an indication of whether the print dimensions need to be scaled up or down to achieve the intended fit, and therefore interlocking objects should be included in the testing. For future design of radiation therapy treatment equipment, this commissioning step provides information on the level of geometric accuracy that should be expected during device quality control (see section 8.5).

At this stage, it may be advisable to evaluate the effects of radiation on the geometry of a sample of test objects [29, 30].

8.3.5 Characterising the physical properties of materials

In addition to external geometric accuracy, various radiation therapy applications may require particular physical properties to be exhibited by 3D-printed objects. Tests of these properties should also be completed at the commissioning stage. For example, use of 3D-printed devices as brachytherapy applicators will require 3D-printed materials to support catheter channels or seed cavities without collapsing, use of 3D-printed devices as patient supports or immobilisation devices will require 3D-printed materials to have a low enough density to not perturb the radiation beam while having a high enough strength for patient safety (see chapter 9), use of 3D-printed devices as radiation therapy bolus may require materials to be flexible enough for comfort while remaining rigid enough to ensure a reproducible setup [31].

8.3.6 Characterising the density properties of materials

As noted in section 8.2.4, the ubiquity of kV CT imaging systems in radiation therapy centres has meant that it is common for material densities and RED values to be derived from HU numbers from kV CT scans. Therefore, it is important during 3D-print service commissioning to establish the validity of this method with respect to the materials used for 3D printing. This important verification check can be completed by performing transmission, tissue-maximum ratio (TMR), and/or percentage depth dose (PDD) measurements with the printed material in question and either converting the results into effective RED values for comparison with kV CT derived results [24] or using the radiation therapy treatment planning system to calculate TMRs or PDDs using kV CT scans of the material and testing these calculations against the physical measurements [32]. Alternatively, some radiation therapy treatment units are able to produce MV CT scans that can also be used to validate the kV CT results [17].

Once the kV CT has been validated for use for this purpose, kV CT imaging can be used to derive material densities and RED values for a range of test objects printed at a range of infill densities [33], cell size [34], or mixture rates of different materials [35]. This information can then be used to derive calibration curves in the form infill density (or other parameter) vs RED for each material, so that if there is a need to 3D print a material with a particular RED in the future (say, to match an existing treatment device), this calibration will provide the value of infill density (or other parameter) needed to achieve that RED.

Note that further testing of density and other properties (e.g. stopping power and effect on spot size), including the identification of relevant density overrides for treatment planning, may be necessary for electron and other particle beams (e.g. proton, helium or carbon-ion beams) [32, 36–41].

8.3.7 Testing challenging geometries and long print jobs

Standard test objects with challenging geometries can be used to evaluate how a 3D-printing system handles narrow structures, unsupported overhangs, sharp corners, and smooth surfaces [26–28]. Test printing of such objects using the printing materials and parameters selected via the work described above is helpful in 'stress testing' the 3D-printing system to identify print geometries that might cause future print failures or unacceptable deviations from planned geometry or density. This information can help to inform recommendations regarding 3D-print orientation and overhang support options. This information might also feed back into another iteration of 3D-printing parameter testing, as described earlier.

As noted in chapters 6, 9 and 10, some 3D-printing jobs can take many hours or days to complete. In preparation for the possibility of needing to run print jobs out of hours, overnight, or on weekends, it is very important to establish that the chosen 3D printer does not exhibit thermal runaway, heat creep, or other potentially dangerous issues. Thermal runaway is difficult to test for, consequently information in manuals and user guides should be checked against the published literature to verify that this feature is available and functional for the printer concerned. Heat creep testing can be performed by starting a long print early in the day and then periodically using a thermometer probe to measure and record the temperature at various points on the print head.

3D printers that automatically pause when the material supply is interrupted should also be investigated to identify whether heated elements (e.g. the hot end in an MEX printer) are kept at high temperatures during this pause; and if so, then whether this feature can be easily deactivated (so that the printer continues printing nothing until the job is completed and then cools down).

Test printing of long print jobs by starting early in the day and then frequently reviewing print progress until print completion is an absolute minimum test before a 3D printer may be trusted to run without staff present in the radiation therapy centre.

8.3.8 Evaluating 3D-print reproducibility and consistency

Short- and long-term reproducibility are fundamental commissioning tests for many radiation therapy systems. For a given combination of 3D printer, material, and printing parameters, short-term reproducibility can be assessed by repeatedly completing the same print job numerous times and then examining relevant parameters of interest for all the resulting objects. Craft *et al* [20] completed this test by 3D printing 10 cubes in each of four different materials and then evaluating mean HU values from kV CT scans that were repeated over a period of 5 weeks. By this method, they were able to identify the density uncertainties arising from variations in their 3D-printing materials, as well as the (much smaller) changes in the 3D-printed objects in the weeks after their fabrication [20].

It is worth developing a test for long-term reproducibility during the commissioning period so that this test can be used for periodic quality control of the printer, material, and printing parameters after the clinical release of the 3D-printing service. A design for a test object that would provide relevant information if used for printer quality control in a radiation therapy bolus 3D-printing service is shown in figure 8.2. Periodic testing and evaluation of a test object such as this would help to identify if 3D-printing performance starts to drift (see section 8.5).

8.3.9 Completion of end-to-end testing

End-to-end testing is very important, especially for clinical services where 3D-printed devices are to be used for patient treatments. End-to-end testing should be used to identify areas of miscommunication and process bottle-necks, in addition to

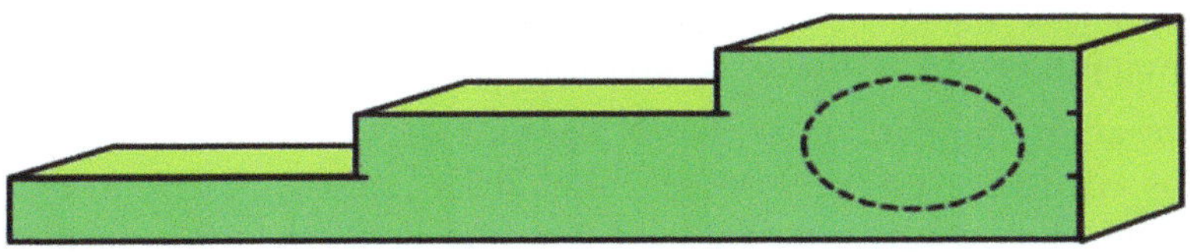

Figure 8.2. Suggested test object for 3D-printing tests as part of on-going quality assurance. Each stepped section is 20 mm wide and 5 mm high. The text object has a maximum width (base) of 60 mm, a maximum height (right-hand side) of 15 mm and a depth (into page) of 30 mm. The dotted line shows a suggested region-of-interest for HU evaluation.

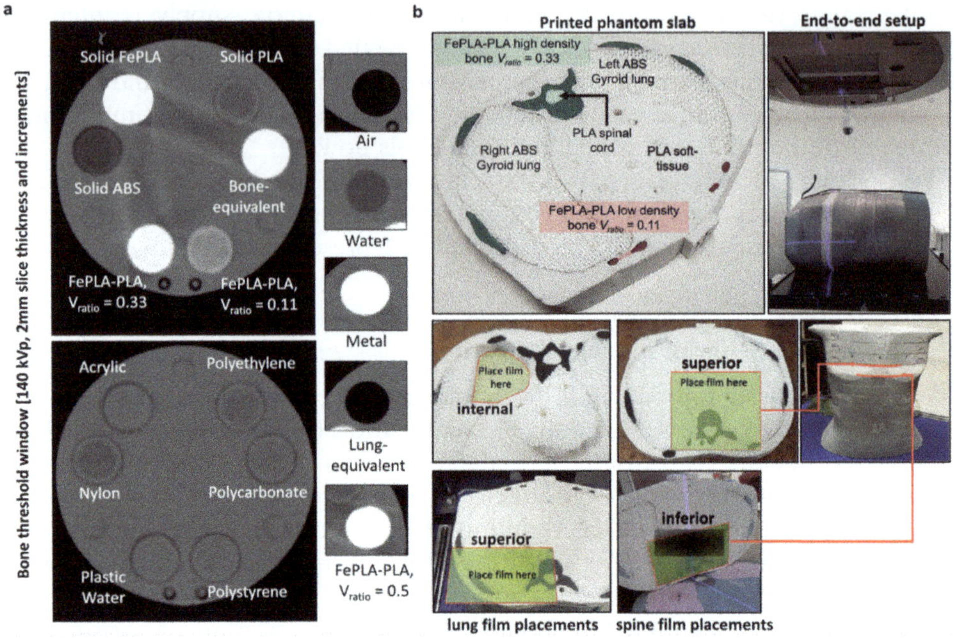

Figure 8.3. Commissioning example. 'Manufactured phantom inserts and phantom slab. (a) CT images of the CT electron density phantom containing commercial, and 3D-printed phantom inserts, (b) end-to-end setup of the CIRS phantom with the 3D-printed phantom. Radiochromic film placements on the 3D-printed phantom, which include two films for spine stereotactic ablative body radiation therapy (SABR; superior and inferior spine regions) and two films for lung SABR (internal and superior lung sub-section, L7)' [16]. Reprinted from [16], Copyright (2022) with permission from John Wiley & Sons.

any issues with the 3D-printed devices themselves. End-to-end testing should include the entire proposed workflow to assess its effect, including requesting the 3D-printed device, communicating of (dummy) patient information and device design constraints, acquiring relevant images (e.g. phantom CT scan), designing, preparing, printing, and post-processing the 3D-printed device, and then using the device for its intended purpose, including all medical imaging involved at treatment and preferably including a dose measurement with and without the 3D-printed device. An example of an end-to-end test of a 3D-printed phantom is shown in figure 8.3

8.3.10 Development of a 3D-print sanitisation process

The need to clean, disinfect or sterilise 3D-printed devices used for radiation therapy applications will depend on the purpose and mode of use of each device (e.g. in contact with a patient or not, used internally or not, and used for one patient or multiple patients), as well as published information on material safety, national standards and regulations. Guidance in this area should be sought from relevant regulators. However, if sterilisation of a particular device is required, then evaluating the potential effects of the sterilisation process on the 3D-printed device will become an important step in commissioning the 3D-printing service. This testing is important because some nominally sterilisable materials are detrimentally affected by sterilisation [42] (see chapter 9). Once an appropriate sanitisation process has been identified, then documentation of that process must also be completed during commissioning.

8.3.11 Planning of routine maintenance for a 3D printer

Routine maintenance of 3D printers is addressed in section 8.4, but it should be noted that the identification of required routine maintenance actions and their frequency, and the setting up of a schedule and log for the performance of those actions, should be completed and documented as part of the commissioning process.

8.3.12 Development of quality control processes

The identification of required quality control actions, their frequency, their initial tolerances, and setting up of a schedule and log for performance of those actions should be completed and documented as part of the commissioning process. For example, the test object developed as part of the 'consistency and reproducibility' tests described above (see figure 8.2) could be periodically reprinted and reevaluated as part of 3D-printer quality assurance. The short-term reproducibility tests completed during commissioning will provide guidance regarding the selection of initial tolerances for this test. The tests, frequency, and initial tolerances for quality control of individual 3D-printed radiation therapy devices should also be identified during commissioning, although this is described in more detail in section 8.5.

8.3.13 Preparation of a commissioning report

The results of the commissioning work described in this section should be recorded in a final commissioning report. This report should document commissioning outcomes, including all of the tested and approved 3D printers and materials and all of the 3D-printing parameters that are recommended when using those printers and materials. The commissioning report should record the methods by which the printers, materials, and parameters were tested and summarise the outcomes of those tests.

The report should relate commissioning outcomes to potential hazards (in terms of detrimental effects) explicitly record all identified 3D printing limitations and restrictions relating to the equipment tested (e.g. this material cannot be used with this printer, this printer cannot be run overnight, or this material can be used for treatment planning only if the following density override is applied).

The commissioning report should act as a resource, providing troubleshooting advice to improve print quality, and providing test methods and results to guide future 3D printer or material commissioning work. An important role of the commissioning report is to record (or provide links to records of) baseline data for all of the maintenance tests and quality control tests that were recommended by the commissioning process, so that on-going monitoring via quality assurance has appropriate continuity throughout its clinical use.

Finally, the commissioning report must state which 3D printers (including serial numbers) were tested and which materials (including manufacturer, date, and batch numbers) were tested with those printers, as well as who completed the testing (i.e. who to contact for further information) and on what dates the various components of the 3D-printing service were released for clinical use. The commissioning report should be reviewed for clarity by staff not directly involved in the work.

8.3.14 Provision of written instructions

In addition to the commissioning report, it is advisable to produce a set of concise instructional documents, to advise users on the correct (commissioned and clinically released) methods for producing radiation therapy devices within the 3D-printing service. These instructional documents should not replace comprehensive staff training but can help act as reminders of correct practices regarding:

- Which materials are approved for use with which printers;
- Materials safety data sheets for all of the chemicals used in the 3D-printing service and stored in the 3D-printing workspace, and standard operating procedures for handling and reporting incidents attributed to 3D printing (e.g. resin in the eye, abrasion from sanding or grinding tools, etc);
- What 3D-printing parameters should be used for each material-printer combination and for each different type of device to be printed (i.e. to achieve required strength, smoothness, density, etc, while avoiding deformations, stringing, surface irregularity, internal voids, etc);
- Limitations on print size (lateral and longitudinal dimensions to fit within printable area, minimum thickness to achieve density consistency, and maximum thickness to fit within printable area);
- Limitations on print position, orientation, supports, and brims (e.g. guidance in case some areas of the print-build plate are prone to produce more warping than others, and if commissioning tests highlighted a minimum print area that must be in contact with the build plate to avoid detachment, see also chapter 5);
- Additional setup advice that may be required for specific material types (e.g. how to setup high-density filament for successful printing or how to avoid common dual-head printing issues);
- Post-processing advice (e.g. to maximise smoothness of prints placed in contact with sensitive patient anatomy);
- Cleaning and sterilisation advice (i.e. what procedures to use with what types of prints)
- Tips for troubleshooting common issues (e.g. print not adhering to build plate, print overly difficult to remove from build plate, print head clogged, etc);
- Instructions for pausing and resuming prints in situations where printing material is about to run out;
- Instructions for changing printing materials (e.g. instructions for replacing filaments for MEX);
- Instructions for handling materials (e.g. MEX filaments to be kept in air-tight containers, VPP materials to be kept in light-tight storage);
- Instructions for completing 3D-printer maintenance;
- Instructions for completing 3D-printer quality control testing; and
- Instructions for completing 3D-printed device quality control testing.

8.3.15 Creating logs

As noted in chapter 7, it is important to create, share, and regularly review logs of 3D-printing activity, 3D-printer maintenance and quality control, 3D-print quality control, and sterilisation processes, as well as 3D-printing faults and complaints, as part of the overall 3D-printing quality management system. These logs should be created as part of commissioning, and their existence and locations should be recorded in the commissioning report.

8.3.16 Provision of staff training

The results of testing performed during commissioning can inform the content of a comprehensive programme of staff training. For example, commissioning tests will demonstrate how things can 'go wrong', by evaluating the effects of varying 3D-printing materials and settings. Staff who will be involved in creating 3D-printing requests, designing 3D-printed devices, running the 3D-printers and post-processing equipment, completing checks and quality control measurements of the printed devices, and using the devices clinically should all receive a level of training that is appropriate for their role. For example, staff involved in running the 3D printers should be trained in 3D-printer safety, plus how to start print jobs, observe for early issues, change materials, perform routine maintenance, and perform basic trouble-shooting tasks, whereas staff whose only role might be requesting a 3D-printed device or using it in treatment might only require a short presentation on the advantages and limitations of 3D printing. All involved staff should be directed to use the relevant logs and instructed on how to report faults.

By providing comprehensive staff training for 3D-printing processes, as well as the clinical use and quality control of 3D prints, radiation therapy centres can minimise the likelihood of unsatisfactory results being produced by human error (see example in figure 8.4)

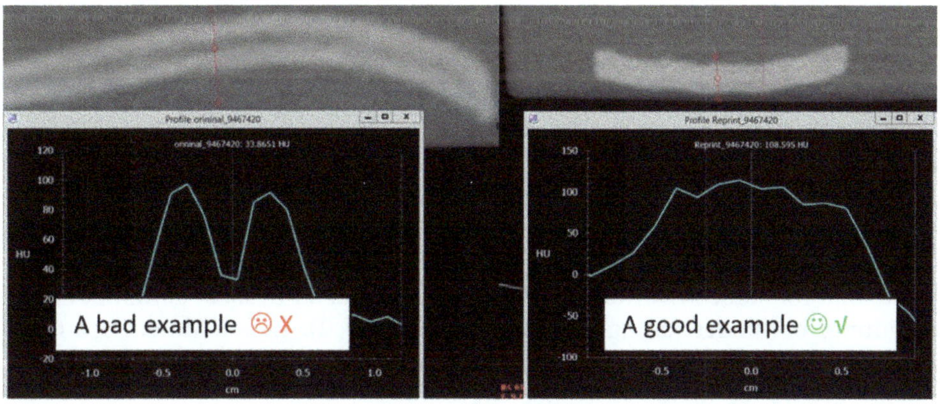

Figure 8.4. Example of failure mode, highlighting the importance of commissioning and documented print process: a pair of good and bad examples for a 1 cm bolus with and without internal air gap, respectively.

8.3.17 Repetition of commissioning for new equipment

The radiation therapy 3D-printing commissioning tasks that are described throughout the rest of section 8.3 (or the commissioning tasks identified as necessary by each radiation therapy centre via their local risk assessment) should be revisited each time that new 3D-printing equipment, processes, or material types are introduced into the service. Different 3D printers can produce different results, due to obvious differences in design—even two printers of exactly the same make and model can produce different results as a result of subtle differences in their calibration parameters [21]. Different 3D-printing materials also behave differently when processed with the same 3D-printing settings (e.g. flexible MEX filaments are notoriously more difficult to work with than non-flexible filaments [43, 44]), and so it is inadvisable to change material types without completing at least a subset of the tests suggested here.

Fortunately, the commissioning report and associated measurement records, in addition to the institutional experience gained by completing this process once, will make subsequent repetitions of the commissioning process easier and more efficient.

8.4 3D-printer maintenance

If a 3D printer is to be used to fabricate devices for radiation therapy use, then the maintenance steps that are needed in order to keep the printer performing accurately and consistently should be addressed as part of the overall quality assurance process. There is a wealth of published information (including online video tutorials) regarding 3D-printer maintenance. Therefore, this section is intended to act as a reminder to include planning for maintenance in the 3D-printer commissioning process and to include completion of maintenance as part of 3D-printing quality assurance, rather than as detailed guidance on how to do printer maintenance.

Clearly, the most useful first step towards developing the necessary programme of routine maintenance is to review published maintenance advice for the particular printer or printer type, including instructions or provided by the manufacturer. Generally, routine maintenance is likely to consist of cleaning and dusting, checking and replacing parts that can be easily worn or blocked, tightening screws, lubricating mechanisms, and performing simple performance checks.

For example, for MEX printers, it is common for routine maintenance to include cleaning surfaces and guide tubes, checking belt tensions, tightening isolators and lubricating drive screws, as well as periodically checking the levelling of the print-build plate. After firmware updates or any changes in filament type (especially, for example, a change from a rigid to a flexible filament), it is also advisable to check the extrusion rate calibration to avoid irregular surfaces or dramatic changes in print density [21] (see section 8.2.1).

Note that for some 3D printers (including high-end VPP, sheet lamination, multi-jet printing, or binder jetting printers, see chapter 2), user maintenance may be inadvisable due to physical, mechanical and/or electrical safety hazards, and may be explicitly prohibited under warrantee conditions.

A simple performance check that can be performed periodically, whether or not any other forms of maintenance are completed, is to evaluate the result of printing a standard test object selected from a library of publicly available stereolithography (STL) models, such as '3Dbenchy' [26] (see figure 3.1) and 'All-in-one' [27]. Alternatively, it may be useful to design of an in-house test object to eliminate irrelevant variables and encourage examination of the specific features of interest for the intended radiation therapy application (see figure 8.2).

The required frequency of routine 3D-printer maintenance will depend on the demands of the particular radiation therapy centre in which it is used. For instance, more frequent maintenance will be required on 3D printers that are used intensively, although function tests may be required prior to every use of a printer that stands idle for weeks at a time. The maintenance schedule will also be affected by how rigidly the use of the 3D printer is controlled—whether use is open to all staff or limited to a small number of specifically trained individuals, and whether usage is indiscriminate or limited to selected materials and purposes.

Once decided, the maintenance actions and schedule should be documented, together with their rationale and references to any sources of recommended actions. Staff involved in 3D printing should be trained in the required maintenance actions, or advised not to undertake maintenance actions but instead call a representative from the manufacturer if necessary.

Maintenance logs and systems for tracking outcomes of maintenance checks and usage of spare parts should be setup and shared with relevant staff, so that maintenance actions can be tracked as part of quality management (see chapter 7) and any recurring 3D-printer maintenance problems can be identified and resolved.

8.5 3D-print quality control

Having established a framework for the commissioning stages of the quality assurance process in section 8.3, this section provides guidance regarding suggested quality control steps through the example of 3D-printed patient-specific radiation beam-modifying devices, such as bolus, compensators, and range shifters. Beam-modifying devices (which are referred to using the general term 'bolus' throughout this section for brevity) are taken as the focus of this section due to their obvious and direct effect on the radiation therapy treatment dose delivered to patients. However, the guidance provided in this section should be broadly applicable across the range of 3D-printed radiation therapy devices. (See also chapter 9 for additional considerations relating to devices used with patients and chapter 10 for additional considerations relating to devices used for radiation therapy system and dose verification.)

Every potential user of 3D-printed beam-modifying devices should understand that the exact compositions and characteristics of filaments from different vendors can vary and that these characteristics can also change with time [20].

Even after a radiation therapy centre has setup a bolus 3D-printing service, tested and commissioned their 3D printer(s), and selected their filament and printing parameters, unexpected changes in filament composition, unusual bolus geometries, or human error can produce printed boluses that deviate from the densities and geometries, and therefore dosimetric effects that were initially planned for those boluses. Even boluses produced for radiation therapy use by external commercial suppliers can have serious flaws that would make them unsuitable for safe and effective clinical use [19]. For all of these reasons, a stringent process of quality control is necessary within the radiation therapy centre at which the bolus is used.

Some additional testing required for charged particle beams are addressed in chapter 9, but generally, for all radiation sources, the key question that quality control testing should answer is: can the radiation therapy treatment be delivered as planned if this device is used? The approach to this answering question will depend on clinical workflows, as well as the extent of prior testing of material samples that has been completed.

For clinical workflows where the 3D-printed device is included in the treatment planning CT of the patient, the key area of investigation is whether or not the 3D-printed device is modelled accurately in the radiation therapy treatment planning system. First, it must be established that a kV planning CT of the bolus material accurately represents that material's effect on an MV treatment beam [24, 29]. If that is not the case, then an appropriate density override needs to be identified for use in treatment planning (see description of work from Burleson *et al* [32] in chapter 9). Second, the accuracy of the treatment planning system's calculations of dose in the presence of air gaps and density variations must be investigated. This investigation can inform the selection of tolerances for use in quality control [45].

For clinical workflows where the 3D-printed device is not included in the treatment planning CT of the patient, additional verification is required. In these workflows, a virtual device is designed within the treatment planning system (or other software) and used for the treatment dose calculation. Therefore, it is necessary to establish that the printed device would have the same effect on the treatment beam as that virtual device. Specifically, there are two key variables that need to be tested: the print geometry and the print electron density. (Additional tests for proton range shifters are briefly summarised in chapter 9.)

For this second workflow, the most widely-reported quality control process involves acquiring a separate CT scan of the printed device, and evaluating the device's geometry and density using that scan [1, 32, 45, 46]. Obviously, this method also requires that the suitability of kV CT scanning be established in advance for the print material under investigation [24, 29].

The quality control CT scan must be completed at a high resolution using the smallest usable field of view and the narrowest commissioned slice thickness [45]. It is also recommended that the 3D-printed device be placed on top of a radio-transparent stand, so that the scanning field of view avoids inclusion of the CT couch or any other materials that might confound the density assessment [45] (see chapter 4). Open-source software such as 3D Slicer can then be used to segment the device and produce a HU histogram, from which a RED histogram and a mean RED calculation can be obtained [1, 46]. A new STL of the 3D-printed device can be created from the segmented surface for geometric comparison against the STL of the virtual bolus produced by the treatment planning system (or other software) [1, 46]. An example document summarising this testing and its possible outcomes is shown in figure 8.5.

QC Test	Details	Results/Deviations
T2- Visual and manual inspection	**Inspection:** The device will be lifted and turned to enable 360° visual inspection. The device will be manually stressed (e.g. squeezed in hands) and checked for smoothness. **Acceptance criteria:** 1. Printed device matches device specification. 2. No visible print defects (e.g. holes >3mm) and no residual supporting structure. 3. No noticeable loss of structural integrity. 4. Suitable smoothness.	1. ☐ Pass: Device matches specifications in Case Plan ☐ Deviation: Device deviated from specification in Case Plan. Specify deviations: 2. ☐ Pass: There are no visible print defects or residual support structures present. ☐ Deviation: Device has visible print defects or residual support structures. Specify defects: 3. ☐ Pass: Device has no visible cracking nor loss of structural integrity after manual inspection ☐ Deviation: Specify deviations: 4. ☐ Pass: Device is of suitable smoothness ☐ Deviation: Device is not of suitable smoothness. Specify deviations:
T3- Dimensional inspection	**Inspection:** Device scanned (e.g. CT or laser or structured light), and scan compared against initial design via Hausdorff distance metric. **Acceptance criteria:** • As specified by medical professional OR • Within image modality uncertainty (e.g. <0.5 mm) for 95% of volume AND <2 mm over entire volume.	☐ Pass: Device passed dimensional inspection. See details below. ☐ Deviation: Specify deviations: 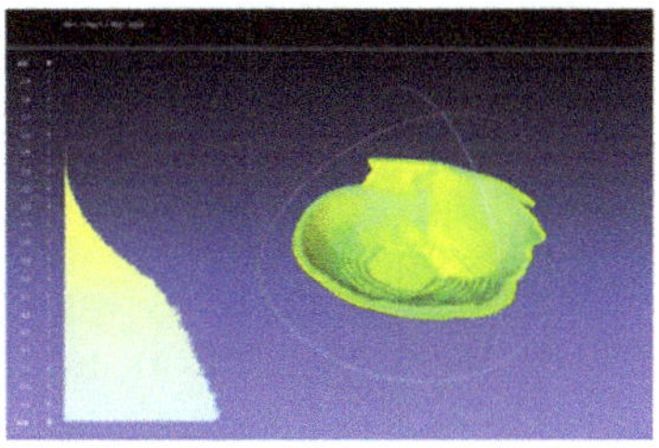
T4- Radiological inspection	**Inspection:** Sampling of CT image pixel values within body of device, conversion to density and/or electron density using imaging system specific function. **Acceptance criteria:** • As specified by medical professional OR • Within image modality uncertainty and standard deviation in mean density; mean within 0.1 of value requested, standard deviation in mean density <0.2.	☐ Pass: Device passed radiological density inspection. See details below. ☐ Deviation: Specify deviations: 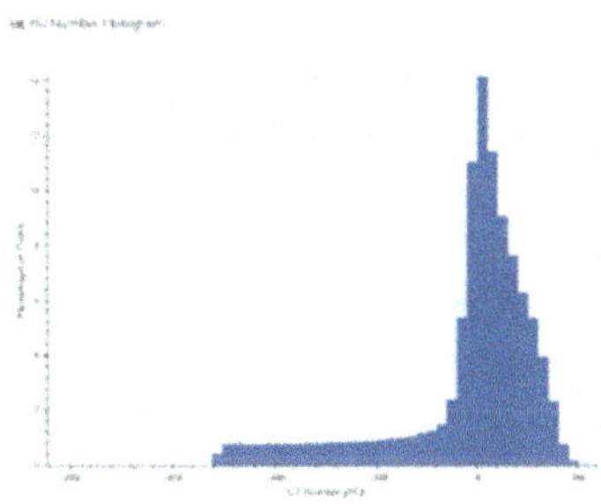

Figure 8.5. Example quality management document summarising quality control steps and possible outcomes for a 3D-printed bolus. *Not for uncritical use—all tolerances should be reevaluated for suitability under local conditions prior to adoption.*

The current lack of published guidelines or codes of practice relating to 3D-print quality control is a serious impediment to the systematic adoption of quality control tests and tolerances for 3D-printed radiation therapy devices. In the absence of such agreed standards (while, for example, recommendations from AAPM Task Group 336 are currently in preparation), tolerances for these tests must be established locally with reference to the dose calculation accuracy of the treatment planning system and the dosimetric effects of errors in device fabrication. Previous work by

Kairn *et al* [45] shows an example of how this can be done by using physical measurements to establish the accuracy of one treatment planning system in the presence of air gaps and density variations, and then using that treatment planning system to evaluate the dosimetric effects of a range of 3D-printed device fabrication errors. Note that this study [45] was focussed on one treatment type (VMAT) for one anatomical site (nose), so resulting tolerances should not be broadly applied without further investigation.

Routine quality control testing of printers and filaments is also important, and may eventually replace quality control testing of individual prints in well-established services. For example, at the Peter MacCallum Cancer Centre where approximately 2000 patient-matched 3D-printed devices have so far been produced, quality control of the printing process is performed using monthly reprinting of the test object shown in figure 8.2, which is physically measured and CT scanned to assess thickness, HU, and density. The tolerances for the mean HU assessment from the CT of this object are typically between -50 and 200 HU if 3D prints are to be used for MV photon treatments, and between 0 to 150 HU if 3D prints are to be used for electron treatments, although these can vary with treatment modality, as shown in table 8.2.

Given that the monthly quality control test of the 3D-printing process includes geometric measurements in addition to the HU assessment, and based on clinical experience with hundreds of 3D-printed treatment devices, the established 3D-printing service at the Peter MacCallum Cancer Centre does not include geometric assessments when evaluating each individual 3D-printed bolus. Rather, a simplified quality control test is performed for each patient-specific 3D-printed bolus consisting of a CT scan and mean HU check. If the HU of the bolus differs from nominal (usually 0 HU, for water-equivalence), then the checking physicist informs the clinical teams detailing the mean HU recorded at CT and any action required. The new value can then be amended in the treatment plan and the plan recalculated prior to the patient's treatment.

Hypothetically, this quality control checking of 3D-printed patient-specific bolus could be reduced even further by radiation therapy centres with sufficient experience to inform reliable risk assessments; that is, the frequency of print quality control failures needs to be known and the potential severity of effects from undetected failures needs to be understood before the risks of reducing quality control checking can be predicted. In such cases, if regular quality control of the printing process is being completed via evaluations of test objects as described above,then it may be possible to follow a streamlined quality control process where testing of patient boluses is reduced to spot checking one in every ten boluses (for instance) with reference to local requirements (see example in table 8.2) within the first three fractions of the corresponding treatments (or within the first 10% of the treatment, if fewer than 20 fractions are to be delivered).

However, any change from quality control of individual devices to quality control of the printer and filament plus spot checking of occasional boluses should only be considered with reference to a risk assessment that takes into account the likelihood and frequency of print quality changes due to variations in the filament or printer, as

Table 8.2. Example bolus quality control requirements. *Not for uncritical use—all tolerances should be reevaluated for suitability under local conditions prior to adoption.*

Modality	Mean HU	HU variation	Thickness accuracy	Minimum thickness	Other
MV photon 3DCRT	Within 100 HU of nominal	Within 100 HU of mean	± 1 mm	5 mm	–
MV photon IMRT/VMAT	Within 100 HU of nominal	Within 100 HU of mean	± 1 mm	5 mm	Air gaps to skin < 2 mm
Electron	Within 50 HU of nominal	Within 50 HU of mean	± 0.5 mm	5 mm	Air gaps to skin < 2 mm
kV photon	Within 50 HU of nominal	Within 50 HU of mean	± 0.5 mm	5 mm	Air gaps to skin < 1 mm
Brachytherapy	Within 50 HU of nominal	Within 50 HU of mean	± 1 mm	5 mm	Distance of catheters to skin within 1 mm of planned
SGRT	Per treatment modality	As per modality	As per modality	5 mm	Surface reflection resembles human skin

well as human error. Such a programme would also require a willingness to return to testing of individual devices when changing to a new printer or filament, when encountering particularly unusual print designs and when expanding the service to include types of devices that were not previously included in the 3D-printing service.

All 3D-printed device quality control should be documented as part of the broader 3D-printing quality management system (see chapter 7) and quality control test results for individual patients should be included with other treatment records. The 3D-printed patient-specific beam-modifying device quality management system should also include agreed and documented processes for requesting, designing, printing, verifying, using, and sanitising the prints. Written instructions should also be provided for any new processes that are not part of the standard radiation therapy treatment procedure. Such instructions may include how to create the bolus design, what density to assign, and which print options to use with which filaments.

Finally, reviews of quality control results should be included in a programme of regular auditing of the process of 3D-printing patient-specific beam-modifying devices [46]. These reviews provide the opportunity to identify trends or step-changes in quality control results, which may highlight specific issues with the printer or filament or may indicate the need for additional staff training. Aggregating and reviewing test results also provides an opportunity to examine and possibly tighten the tolerances used for quality control of 3D-printed devices (see chapter 7).

8.6 Key points

- Quality assurance focusses on processes and aims to avoid defects, whereas quality control focusses on outputs and aims to identify defects; quality assurance includes quality control.
- The requirements of 3D printing in the radiation therapy context are different from (often more stringent than) the requirements of 3D printing in other areas, and so carefully planned and documented quality assurance processes are needed.
- Common 3D-printing defects that have appreciable effects in the radiation therapy context include unintended irregular and rough surfaces, geometric errors, density variations, unsuitable doping, bulk deformation and clearance issues, problems with consistency and reproducibility, and print failure.
- Based on the common defects affecting 3D printing in radiation therapy and the clinical consequences of those defects, a risk assessment should be completed to aid in the planning of a local commissioning process that will minimise the likelihood of these deficits affecting a patient, and reduce all risk levels to 'low' or 'negligible.'
- Key commissioning steps for a radiation therapy 3D-printing service include: familiarisation with existing documentation; determination of 3D-printing parameters; evaluation of geometric accuracy; characterisation of physical properties of 3D-printed materials; characterisation of density and HU properties of 3D-printed materials; testing of challenging geometries and long print jobs; evaluation of 3D-print reproducibility and consistency;

completion of end-to-end testing; development of 3D-print sanitisation process; planning of 3D-printer routine maintenance; development of 3D-printer and 3D-print quality control testing; preparation of a commissioning report; provision of written instructions; creation of logs for 3D-printing activity, 3D-printer maintenance, quality control testing, sterilisation processes, and 3D-printing faults; and provision of staff training.

- Routine 3D-printer maintenance should be included as part of the quality assurance process, with manufacturer instructions and published advice being the first step towards initiating that maintenance.
- 3D-print quality control is an important aspect of 3D-print process quality assurance.
- Routine quality control testing is necessary for in-house 3D prints because unexpected variations in materials, changes in printer performance, or human errors can cause print defects with serious consequences for radiation therapy treatments.
- Quality control testing of 3D prints is necessary, even when 3D prints are supplied by external and commercial suppliers.
- In the absence of relevant published guidelines or codes of practice, quality control tests and tolerances should be defined with reference to the purpose of each type of 3D print and the variables that would affect its use (e.g. RED and geometry for bolus.)
- Commissioning tests and results, 3D-printing instructions, maintenance logs, and quality control test procedures and results should be documented and regularly reviewed as part of the broader clinical 3D-printing quality management process (see chapter 7).

References

[1] Sasaki D K, McGeachy P, Aviles J E A, McCurdy B, Koul R and Dubey A 2019 A modern mold room: Meshing 3D surface scanning, digital design, and 3D printing with bolus fabrication *J. Appl. Clin. Med. Phys.* **20** 78–85

[2] Zhao Y, Moran K, Yewondwossen M, Allan J, Clarke S, Rajaraman M, Wilke D, Joseph P and Robar J L 2017 Clinical applications of 3-dimensional printing in radiation therapy *Med. Dosim.* **42** 150–5

[3] Sharma A *et al* 2018 Low-cost optical scanner and 3-dimensional printing technology to create lead shielding for radiation therapy of facial skin cancer: first clinical case series *Adv. Radiat. Oncol.* **3** 288–96

[4] Cheng J C, Dubey A, Beck J, Sasaki D, Leylek A and Rathod S 2022 Optical scan and 3D printing guided radiation therapy–an application and provincial experience in cutaneous nasal carcinoma *3D Print. Med.* **8** 8

[5] Briggs M, Clements H, Wynne N, Rennie A and Kellett D 2016 3D printed facial laser scans for the production of localised radiotherapy treatment masks–a case study *J. Vis. Commun. Med.* **39** 99–104

[6] Crowe S *et al* 2022 Evaluation of optical 3D scanning system for radiotherapy use *J. Med. Radiat. Sci.* **69** 218–26

[7] Asfia A, Novak J I, Mohammed M I, Rolfe B and Kron T 2020 A review of 3D printed patient specific immobilisation devices in radiotherapy *Phys. Imaging Radiat. Oncol.* **13** 30–5
[8] Miron V M, Etzelstorfer T, Kleiser R, Raffelsberger T, Major Z and Geinitz H 2022 Evaluation of novel 3D-printed and conventional thermoplastic stereotactic high-precision patient fixation masks forradiotherapy *Strahlentherapie und Onkologie* **198** 1032–41
[9] Mattke M, Rath D, Häfner M, Unterhinninghofen R, Sterzing F, Debus J and Giesel F L 2021 Individual 3d-printed fixation masks for radiotherapy: first clinical experiences *Int. J. Comput. Assist. Radiol. Surg.* **16** 1043–9
[10] Wilks R, Crowe S B, Chan P, Cheuk R and Kairn T 2023 A 3D printed patient-specific vaginal mould for brachytherapy *J. Phys.: Conf. Ser.* **2630** 012026
[11] Laan R C, Nout R A, Dankelman J and van de Berg N J 2019 MRI-driven design of customised 3D printed gynaecological brachytherapy applicators with curved needle channels *3D Print. Med.* **5** 1–8
[12] Wiebe E, Easton H, Thomas G, Barbera L, D'Alimonte L and Ravi A 2015 Customized vaginal vault brachytherapy with computed tomography imaging-derived applicator prototyping *Brachytherapy* **14** 380–4
[13] Sim L 2016 Novel application of 3D printing in brachytherapy using MED610 3D printed insert for I-125 ROPES eye plaque *Australas. Phys. Eng. Sci. Med.* **39** 863–70
[14] Simpson-Page E, Hamlett L, Lew D, Stephens H, Wilks R, Kairn T and Crowe S B 2021 3d printed brachytherapy jig for Reference Air Kerma Rate calibration *Phys. Eng. Sci. Med.* **44** 1141–50
[15] Livingstone A G 2022 3d printed PTW TRUFIX-compatible detector holders *Phys. Eng. Sci. Med.* **45** 356
[16] Tino R B, Yeo A U, Brandt M, Leary M and Kron T 2022 A customizable anthropomorphic phantom for dosimetric verification of 3D-printed lung, tissue, and bone density materials *Med. Phys.* **49** 52–69
[17] Kairn T, Zahrani M, Cassim N, Livingstone A, Charles P and Crowe S 2020 Quasi-simultaneous 3D printing of muscle-, lung-and bone-equivalent media: a proof-of-concept study *Phys. Eng. Sci. Med.* **43** 701–10
[18] How S, Banjade D, Crowe S, Dillon G and Skimmings A 2023 An anthropomorphic 3D printed inhomogeneity thorax phantom slab for SBRT commissioning and quality assurance *Phys. Eng. Sci. Med.* **46** 575–83
[19] Brown K, Kupfer T, Harris B, Penso S, Khor R and Moseshvili E 2022 Not all 3D-printed bolus is created equal: Variation between 3D-printed polylactic acid (PLA) bolus samples sourced from external manufacturers *J. Med. Radiat. Sci.* **69** 348–56
[20] Craft D F, Kry S F, Balter P, Salehpour M, Woodward W and Howell R M 2018 Material matters: analysis of density uncertainty in 3D printing and its consequences for radiation oncology *Med. Phys.* **45** 1614–21
[21] Charles P, Kairn T and Crowe S 2022 An unexpected change in the density of 3D printed PLA: On the importance of e-step calibration and QA to the clinical workflow *Phys. Eng. Sci. Med.* **45** 377
[22] Crowe S B, Bennett J, Lathouras M, Lancaster C M, Sylvander S R, Chua B, Bettington C S, Lin C Y and Kairn T 2020 Impact of radiopacified bone cement on radiotherapy dose calculation *Phys. Imaging Radiat. Oncol.* **14** 12–6
[23] Tino R, Yeo A, Leary M, Brandt M and Kron T 2019 A systematic review on 3D-printed imaging and dosimetry phantoms in radiation therapy *Technol. Cancer Res. Treat.* **18** 1533033819870208

[24] Dancewicz O, Sylvander S, Markwell T, Crowe S and Trapp J 2017 Radiological properties of 3D printed materials in kilovoltage and megavoltage photon beams *Phys. Med.* **38** 111–8
[25] Hill R, Holloway L and Baldock C 2005 A dosimetric evaluation of water equivalent phantoms for kilovoltage x-ray beams *Phys. Med. Biol.* **50** N331–44
[26] Creative Tools 2015 *#3dbenchy—The Jolly 3D Printing Torture-Test by creativetools.se* [Online] Available: https://www.thingiverse.com/thing:763622 (visited on 05/30/2023)
[27] Trpkoš M 2017 *All in One 3D Printer Test* [Online] Available: https://www.thingiverse.com/thing:2656594 (visited on 05/30/2023)
[28] Gaa Z 2017 *Smart Compact Temperature Calibration Tower* [Online] Available: https://www.thingiverse.com/thing:2729076 (visited on 05/30/2023)
[29] Van der Walt M, Crabtree T and Albantow C 2019 PLA as a suitable 3D printing thermoplastic for use in external beam radiotherapy *Australas. Phys. Eng. Sci. Med.* **42** 1165–76
[30] Kairn T, Crowe S and Markwell T 2015 Use of 3D printed materials as tissue-equivalent phantoms *World Congress on Medical Physics and Biomedical Engineering, June 7–12, 2015, Toronto, Canada* (Berlin: Springer) pp 728–31
[31] Chiu T, Tan J, Brenner M, Gu X, Yang M, Westover K, Strom T, Sher D, Jiang S and Zhao B 2018 Three-dimensional printer-aided casting of soft, custom silicone boluses (SCSBs) for head and neck radiation therapy *Pract. Radiat. Oncol.* **8** e167–74
[32] Burleson S, Baker J, Hsia A T and Xu Z 2015 Use of 3D printers to create a patient-specific 3D bolus for external beam therapy *J. Appl. Clin. Med. Phys.* **16** 166–78
[33] Leary M, Tino R, Keller C, Franich R, Yeo A, Lonski P, Kyriakou E, Kron T and Brandt M 2020 Additive manufacture of lung equivalent anthropomorphic phantoms: a method to control hounsfield number utilizing partial volume effect *J. Eng. Sci. Med. Diagn. Ther* **3** 011001
[34] Tino R, Leary M, Yeo A, Brandt M and Kron T 2019 Gyroid structures for 3D-printed heterogeneous radiotherapy phantoms *Phys. Med. Biol.* **64** 21NT05
[35] Tino R, Yeo A, Brandt M, Leary M and Kron T 2021 The interlace deposition method of bone equivalent material extrusion 3D printing for imaging in radiotherapy *Mater. Des.* **199** 109439
[36] Lindsay C, Kumlin J, Jirasek A, Lee R, Martinez D M, Schaffer P and Hoehr C 2015 3D printed plastics for beam modulation in proton therapy *Phys. Med. Biol.* **60** N231–40
[37] Ju S G, Kim M K, Hong C-S, Kim J S, Han Y, Choi D H, Shin D and Lee S B 2014 New technique for developing a proton range compensator with use of a 3-dimensional printer *Int. J. Radiat. Oncol. Biol. Phys.* **88** 453–8
[38] Zou W *et al* 2015 Potential of 3D printing technologies for fabrication of electron bolus and proton compensators *J. Appl. Clin. Med. Phys.* **16** 90–8
[39] Shirey R J and Wu H T 2018 Quantifying the effect of air gap, depth, and range shifter thickness on TPS dosimetric accuracy in superficial pbs proton therapy *J. Appl. Clin. Med. Phys.* **19** 164–73
[40] Michiels S *et al* 2016 Towards 3D printed multifunctional immobilization for proton therapy: initial materials characterization *Med. Phys.* **43** 5392–402
[41] Shen J, Liu W, Anand A, Stoker J B, Ding X, Fatyga M, Herman M G and Bues M 2015 Impact of range shifter material on proton pencil beam spot characteristics *Med. Phys.* **42** 1335–40

[42] Wilks R, Skewes J, Forrestal D, Kairn T and Crowe S 2023 Evaluating 3D printing materials for intracavitary gynaecological brachytherapy *Phys. Eng. Sci. Med.* **46** 482

[43] Ehler E, Sterling D, Dusenbery K and Lawrence J 2018 Workload implications for clinic workflow with implementation of three-dimensional printed customized bolus for radiation therapy: a pilot study *PLOS One* **13** e0204944

[44] Kairn T, Poroa T, Cleland S, Dawes J, Kenny L, Lancaster C, Lin C and Crowe S 2023 Clinical use of 3D printed intra-oral stents: opportunities and obstacles *Phys. Eng. Sci. Med.* **46** 480–1

[45] Kairn T *et al* 2021 Determining tolerance levels for quality assurance of 3D printed bolus for modulated arc radiotherapy of the nose *Phys. Eng. Sci. Med.* **44** 1187–99

[46] Livingstone A, Simpson-Page E, Poroa T and Crowe S 2022 Clinical implementation of a quality assurance program for 3D printed devices *Phys. Eng. Sci. Med.* **45** 310

Chapter 9

Patient treatments

Tanya Kairn, Rachael Wilks and Samuel C Peet

9.1 Introduction

Three-dimensional (3D) printed accessories can be used for many different applications relating to radiation treatments of patients. These applications, discussed in the following sections, range from beam-modifying devices (e.g. bolus, compensators, and range shifters, in section 9.3) to custom shielding (e.g.shields and apertures, or 'cutouts', in section 9.4) patient immobilisation devices (e.g. stents, supports, and positioning equipment, in section 9.5) and brachytherapy applicators (e.g. external moulds, internal applicators, needle templates, and dose calculation concerns, in section 9.6.4). This chapter describes the use of 3D-printed devices in all of these areas, commencing with a discussion of patient safety (i.e. devices for external and internal use, cleaning, and sterilisation, in section 9.2).

9.2 Patient safety

9.2.1 Regulatory and quality management context

In radiation therapy, 3D-printed devices may be used either in direct contact with the patient's skin or within body cavities. They can also be used in situations where they affect the physical positions of anatomical structures and/or situations where they affect the radiation dose. Therefore, it is very important to consider patient safety when implementing a clinical 3D-printing service and to evaluate the safety of any 3D-printed accessories that may be used in the treatment of radiation therapy patients.

An obvious avenue for investigating, establishing, and maintaining the safety of 3D-printed radiation therapy accessories is to follow national or international standards. Processes dictated by legislation for attainment of certification or registration of medical devices can be followed to achieve safer fabrication and use of treatment accessories, whether or not registration is specifically required or sought.

doi:10.1088/978-0-7503-3907-0ch9

As an analogue, compliance with the USA Food and Drug Administration's guidance for medical device software has recently been discussed as a means to produce thoroughly validated software that is fit for purpose and safe to use, whether or not regulatory clearance is needed by the software developers or the software development projects concerned [1]. For 3D-printed radiation treatment accessories, guidance regarding recommended processes for registered devices has been provided by Australia's Therapeutic Goods Administration to support recent changes to the standards governing medical devices in Australia, as described in chapter 7 and mentioned in chapter 11. Even for radiation therapy centres that are not in Australia or New Zealand, or who are not pursuing registration, this information can be used as a framework to set up validation processes for requesting, fabricating, cleaning, use, recording, and reporting 3D-printed radiation therapy devices.

The establishment of such a quality management system is further detailed in chapter 7, and quality assurance explained in chapter 8. Meanwhile, chapter 6 provides information regarding the safety concerns around 3D printing in a clinical environment. This section builds on these previous chapters by briefly summarising the safety issues relating to patient treatments in particular.

9.2.2 External use

3D-printed radiation therapy accessories may be used externally to the patients, with or without direct contact with the skin. Safety standards for 3D-printed radiation therapy accessories, like other radiation therapy accessories, are determined by whether the 3D-printed devices are designed with reference to patient data and each device is intended for use for one patient, or whether the devices are generic or intended for use with multiple patients (see chapter 7).

All 3D-printed radiation therapy accessories intended for external use in or near radiation therapy treatment sources (e.g. brachytherapy radionuclides or external beam radiation therapy treatment systems, see chapter 1) are subject to safety concerns around the intended or unintended effects on the radiation dose from those sources, which may arise from print density, fit, positioning reproducibility, and geometric accuracy. These issues are discussed separately in relation to particular applications in sections 9.3, 9.4, 9.5, and 9.6.

For example, when considering the use of 3D-printed accessories for patient support or immobilisation, additional safety considerations are raised by the risk of physical injury due to printed accessories unexpectedly buckling, bending, or breaking during use, as discussed in section 9.5.2. By contrast, section 9.4.3 highlights the potential for improving patient and staff safety by using 3D-printed materials to supplant the hammering of lead sheets for use as patient-specific shielding of kilovoltage (kV) radiation therapy beams.

For patient safety, 3D prints that are intended for external use in direct contact with skin or skin lesions should be bio-inert, smooth, non-porous, waterproof, and easily cleanable (see section 9.2.4). Strength and robustness are also especially important for 3D-printed patient support and immobilisation equipment (see section 9.5).

9.2.3 Internal use

Potential internal uses of 3D-printed radiation therapy accessories include: ear packing in combination with bolus to achieve a consistent electron radiation therapy dose [2] or photon intensity modulated radiation therapy (IMRT) dose coverage of treatment volumes close to or including the ears [3] (see section 9.3.2); intra-oral positioning stents, to achieve consistent positioning of targeted oral anatomy and minimise dose to adjacent sensitive healthy tissues [4] (see section 9.5.5); and intracavitary brachytherapy applicators to treat gynaecological and oral cancers [5, 6] (see section 9.6.4). In addition to the safety concerns discussed in section 9.2.2, the potential use of 3D-printed accessories for internal applications requires particular attention to be paid to material safety, as well as cleaning and sterilisation.

Given that established packing materials for the external aural canal have included non-sterile wax and liquid water [2, 7], sterilisation is unlikely to be required of 3D prints intended for intra-aural use. The chosen 3D-printing material should nonetheless be easily cleanable and smooth or flexible enough to minimise patient discomfort [3].

For intra-oral and intra-vaginal applications, where the 3D print is likely to be in contact with a mucous membrane, it is necessary for the printing material to be non-toxic. For example, Cleland *et al* [4] selected a particular form of polylactic acid (PLA) that is marketed as ‘food safe’ and compliant with Australian Standard AS2070-1999 10/2011 [8] to fabricate intra-oral positioning stents for head-and-neck radiation therapy. A prototype patient-specific oral brachytherapy applicator created by Lancellotta *et al* [6] and a prototype patient-specific vaginal brachytherapy applicator created by Wiebe *et al* [5] were fabricated using stereolithography (STL) from resins that were characterised as ‘dental’ and ‘medical grade’, respectively.

Uncritical faith in the safety or reliability of materials due to marketing information, manufacturer’s specifications, or even previous publications is inadvisable. Obviously authors and reviewers can make mistakes. For example, an investigation into the sterilisability of PLA for potential use in fabricating surgical equipment [9] incorrectly stated that PLA was ‘Food and Drug Administration approved as a semipermanent dermal filler’, due to confusion with the similarly named but chemically and physically different poly-L-lactic acid (PLLA). Even recognition by regulatory bodies such as the US FDA approval, European CE mark, and Australian Therapeutic Goods Administration is no guarantee that a material will be safe when used in a context outside the application recognised.

The ISO have developed international standards to evaluate the biocompatibility of a medical device based on a risk assessment approach [10] (for information on risk assessments, see chapter 7). Materials that claim biocompatibility with reference to these standards usually undertake their testing procedures based on a geometry similar to the intended end product. For example, a material advertised for dental purposes would have only been tested for small thin products in contact with the mucous membrane. It is therefore recommended to consider how any new devices with novel geometries may differ from the product assumed by the material's manufacturer, and any risks that this may introduce for patient safety.

Any quality management system for 3D prints intended for internal use must include a documented cleaning process, including specific sterilisation steps for any prints that are used on multiple patients or any patient-specific prints that are used surgically, internally, or in contact with mucous membranes or broken skin (see section 9.2.4). The quality management system should also document the toxicity of the printing material and any specific limitations on the use of the material or requirements of the printing process, to limit the effects of the toxicity. For example, a 3D-printed device cannot be regarded as food safe if the printer (especially the nozzle) that was used to create it was previously used with non-food safe materials. Repeated reviews of the literature on material toxicities is also advisable as part of the periodic review of the entire clinical 3D-printing process, so that printing with materials previously considered inert can be speedily abandoned if evidence of toxicity emerges. (For example, a widely accepted dental grade resin [11] was recently found to leach ovotoxic compounds in an *in vitro* study using mouse oocytes [12].)

9.2.4 Cleaning and sterilisation

As an example framework of best practice requirements for medical device cleaning and sterilisation, the Australian Standard AS/NZS 4187:2014 [13] (and its amendments) outlines the legislated reprocessing standards of medical devices in Australia, categorising the medical device according to its intended use. According to the standard, a medical device which is in contact with a mucous membrane is classified as a semi-critical medical device, while device used in a surgical setting is classified as a critical medical device. It is strongly recommended for semi-critical devices, and required for critical devices, that they are sterilised between uses [13].

While there are biocompatible options for thermoplastics, resins, and other 3D-printable materials, their capacity to be cleaned is variable [14]. Moist heat sterilisation (e.g. autoclave) is the preferred method of sterilisation, as stated by the Australian Standards [13]. However, due to the limitation of thermoplastics in a high heat environment [14, 15], this process has not been widely investigated in preference of low temperature chemical sterilisation (e.g. ethylene oxide [5, 16], hydrogen peroxide [17, 18], etc).

Some materials such as resins printed with stereolithography (SLA) technology, or powders printed with selective layer sintering (SLS) technology, are advertised as being compatible with steam sterilisation. Lindegaard *et al* [19], for example, successfully sterilised their 3D-printed brachytherapy needle guide. However, the user must be mindful of the parameters of their 3D-printed medical device and how these may differ from the expectations of the material's manufacturer. Similar to biological considerations discussed in section 9.2.3, any testing by the manufacturer will be based on limited sample geometries. Any variation from these parameters may result in unforeseen problems [20], as exemplified in figure 9.1, so it is recommended to investigate the effects of sterilisation for a representative sample of any new medical device.

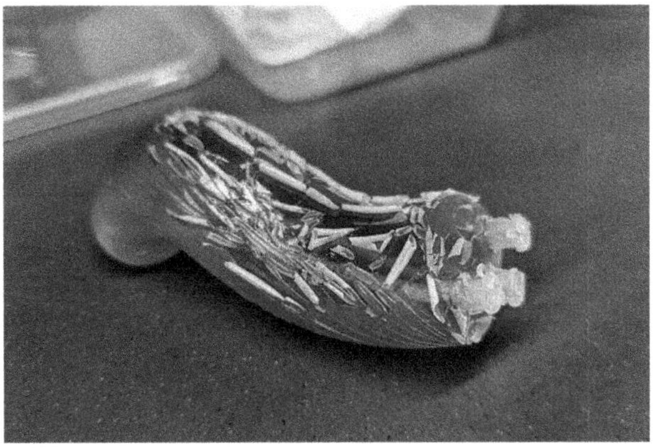

Figure 9.1. Photograph of a photopolymer test object 3D printed using a vat photopolymerisation process for potential use in intracavitary brachytherapy, showing surface flaking after autoclave sterilisation.

Finally, it should be understood that while the creation and adherence to a strict quality control and sanitisation process within a well-documented quality management system is very important for patient safety, the existence of this programme does not guarantee infallibility and does not automatically mean that the clinical 3D-printing service complies with local regulations. Maintaining awareness of local legislation in this area is a necessary aspect of in-house fabrication of clinical equipment, whether using conventional methods or 3D printing.

9.3 Bolus, compensators, and range shifters

9.3.1 Clinical context

Bolus, compensators and range shifters are all used in external beam radiation therapy to modify the depth-dose characteristics of the treatment beam. Bolus is a tissue-similar material that is placed on the patient's skin and used to remove the skin sparing effect of megavoltage (MV) photon and electron beams, as shown in figure 9.2. Compensators may be fabricated from higher density materials and are intended to shape the dose distribution at depth, either to negate the variable beam attenuation from heterogeneous tissues or to spare underlying organs at risk [21–24]. Similarly, range shifters reduce the energy, and therefore the treatment depth of proton beams, helping to achieve coverage of shallow targets [25, 26].

Bolus fabrication is one of the most published clinical applications of 3D printing in the field of radiation oncology [27]. Bolus fabrication was the most common use of 3D printers identified in a 2022 survey of the 112 radiation therapy centres in Australia and New Zealand [28]—12.5% of centres were using in-house systems to 3D-print bolus, with all of these centres printing bolus for photon treatments and half of them also printing bolus for electron treatments [28]. Another 7% of centres reported using in-house systems to 3D-print moulds for bolus fabrication and 30.4% of centres were outsourcing 3D-printed bolus from third parties [28]. Compensator fabrication is a less commonly described application [27], despite being important for conformal passive scattering proton treatments [23, 24] and having potential to

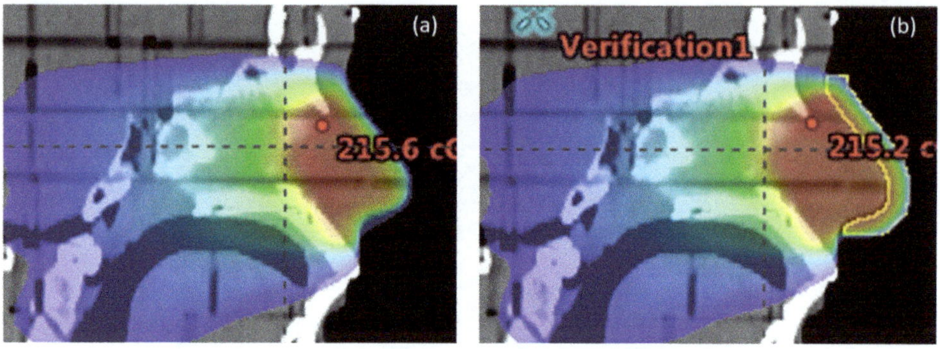

Figure 9.2. Illustration of MV photon bolus effect for a hypothetical squamous cell carcinoma treatment of the nose, showing (a) dose distribution without bolus, where underlying tumour is treated but prescription dose does not reach the skin, and (b) dose distribution for exactly the same treatment plan, with the addition of a bolus (yellow outline), where sparing of underlying tissues is unchanged but prescription dose now covers the whole target including the skin.

aid depth conformity for electron treatments as well [24]. Range shifters are routinely used with passive scattering proton therapy systems [21, 22], although there is growing interest in their use with scanning pencil-beam systems for treatments of shallow targets [25, 26].

When using 3D printing to fabricate patient-specific bolus, compensators, or range shifters, it is important to consider print safety and sanitisation (discussed in section 9.2). The geometry, density, and other diametrically relevant features (see sections 9.3.2 and 9.3.3) of the prints should also be tested and documented as part of the clinical quality management system (see chapters 7 and 8).

9.3.2 Photon radiation therapy

Literature reviews and international surveys suggest that photon bolus production is a major application of 3D printing in radiation therapy clinics [27–29]. Reasons for this include the straightforward process of designing a photon bolus, the suitability of common 3D-printing materials for use in photon beams, and an emerging symbiosis between increasing use of 3D-printed bolus and increasing use of modulated (IMRT or volumetric modulated arc therapy (VMAT)) photon radiation therapy beams in the treatment of superficial tumours and skin cancers.

An ideal photon bolus is usually a uniform-thickness layer of an approximately water-equivalent material that fits the contour of the patient's external surface. This bolus can be challenging to produce and use consistently with methods other than 3D printing [30, 31], but can be easily (even automatically) designed using margin expansions and Boolean operations in a contemporary radiation therapy contouring or treatment planning system [3, 32–39] (see chapter 4).

Material extrusion (MEX) printers (also known as FDM or FFF printers, see chapter 2) are becoming increasingly common in radiation therapy centres [28, 29, 40], and three standard MEX printing materials—acrylonitrile butadiene styrene (ABS), polylactic acid (PLA), and thermoplastic polyurethane (TPU)—have been repeatedly shown to be suitable for photon radiation therapy bolus fabrication [32, 33, 41, 42]. ABS needs to be printed in controlled conditions to minimise safety risks (see chapter 6), but has been shown to provide a better geometric fit with the planned contour of several canthus (corner of eyelid) boluses than paraffin wax [41] despite ABS reportedly deforming when printing larger boluses [33] and tending to expand slightly during print cooling [43]. PLA was the most commonly used 3D-printing material identified in 2022 in Australian and New Zealand radiation therapy centres [28]. PLA is appealing as a bolus material because although it is denser than water, physical measurements in MV imaging and treatment beams have indicated that the density of printed PLA is reliably represented by kV CT scans [33, 42, 44, 45], even when printed with comparatively low infill densities [44]. Use of PLA may also be preferable for environmental reasons (see chapter 6). Flexible filaments such as TPU and soft-PLA may be more challenging to use [34, 46] but have the advantage of retaining a comfortable degree of flexibility when used to fabricate thin bolus components such as nose plugs [32].

Thanks to the ability of modulated radiation therapy (IMRT and VMAT) to achieve complex and concave photon dose distributions that spare neighbouring tissues, interest in using these techniques to treat superficial and skin cancers has been growing over the last decade [47–50]. Achieving superficial dose coverage with MV photon beams obviously requires the use of bolus, and a bolus that fits the patient's external contour produces improved dose distributions compared to a generic bolus that does not fit the patient [3, 51, 52]. For example, clinical trial results published by Anderson *et al* [51] showed lower complication rates from breast radiation therapy when a custom (wax) bolus was used when compared to generic sheet bolus. Anderson *et al* [51] explained that the air gaps that occur when a bolus is not specifically designed to match the patient anatomy results in regions of overdose due to contaminant electrons, as well as regions of skin underdosage due to the creation of a secondary dose buildup region.

This need for a patient-specific photon bolus has led to a parallel growth in the use of 3D printing in radiation therapy centres, alongside the growth in use of modulated MV photon radiation therapy to treat superficial and skin cancers.

For superficial boost volumes, 3D printing has been investigated as an alternative method for achieving the improved breast treatment outcomes described by Anderson *et al* [51] without needing to form patient-specific bolus using the time-consuming and often uncomfortable process of moulding wax, plaster, or other materials directly on the patient's skin [37, 52]. A clinical trial by Zhang *et al* [52] showed that the use of 3D-printed bolus rather than conventional sheet bolus for breast VMAT boosts was associated with improved target dose coverage and homogeneity, as well as reduced dose to the heart and ipsilateral lung, leading to reduced rates of radiation pneumonitis.

For skin cancers in particular, several authors have completed studies of 3D-printed bolus fabrication, fit, reproducibility, and effect on dose distributions for patients with lesions of the extremities, nose, eyelid, ear, scalp, and face [3, 32, 38, 53, 54]. These studies reported a close fit between the 3D-printed bolus and the skin's surface [3, 32, 38, 53, 54], with maximum air gaps of 5 mm or less [3, 32]. Skin cancer patient studies that included comparisons with conventional boluses also reported improved fit and reproducibility, with equivalent or improved dose distributions, when using 3D-printed bolus rather than generic sheets [3, 32].

The advantage of 3D-printed bolus over generic sheet bolus is obvious for the irregular surfaces of the face, ear, and extremities [3, 38, 53]. Several authors have also noted the potential for bolus 3D printing to eliminate some of the dosimetric variability that arises from the manual fabrication of other types of patient-specific bolus, such as wax, rice, or thermoplastic [32, 53, 54], as well as the inter-fraction density variations of wet gauge [31, 32].

Further advantages leading to increased clinical adoption of 3D-printed photon bolus include improved ease and reproducibility of bolus placement [35, 55–57], leading to reductions in patient setup times [37, 57]. The change from manual fabrication of patient-specific bolus to a process of designing and 3D-printing bolus using the patient's planning CT scan also means that the patient does not need to remain present for bolus production. Evidently, the introduction of a clinical 3D-printing service for photon bolus production has the potential to lead to substantial reductions in the amount of time that each patient spends in the radiation therapy centre. Bolus design can also be scripted in the radiation therapy contouring or treatment planning system, and 3D-print jobs can be set up to run out-of-hours and overnight to minimise the bolus fabrication burden on clinical staff.

To maximise 3D-printing efficiency and minimise the severity of issues arising from 3D-printer faults, the printer must have thermal runaway control (see chapters 6 and 8) and should ideally have an auto-pause feature that activates when the filament breaks, snags, or runs out. Careful commissioning tests, including long print runs, must also be completed before the printer is allowed to run without regular observation (see chapter 8). Additionally, the bolus 3D-printing service must include sanitisation (see section 9.2.4), quality control (see chapter 8), and well-documented quality management (see chapter 7).

9.3.3 Electron radiation therapy

Prior to the clinical adoption of 3D printing as a method for fabricating a bolus, the practice of creating a patient-specific bolus with variable thickness was well-established for electron radiation therapy [30]. When using manual dose calculations or early electron dose calculation algorithms, a variable-thickness bolus was needed to fill in and flatten the patient's surface contour for improved electron dose calculation accuracy. In recent decades, there has been growing recognition of the usefulness of variable-thickness bolus for controlling the falloff depth of the election beam by using the bolus as a compensator [2, 58, 59], as exemplified in figure 9.3. For these purposes, a 3D-printed bolus has been proposed as a faster and more adaptable alternative to commercial machinable wax compensator solutions, while also avoiding the discomfort associated with fitting plaster, wax, or thermoplastic directly on the patient's skin [36, 53].

Su *et al* [60] and Zhao *et al* [32] demonstrated the use of a variable-thickness bolus designed using iterative dose calculations to achieve optimised dose distributions, respectively, for a 16 MeV electron treatment of abdominal rhabdomyosarcoma and

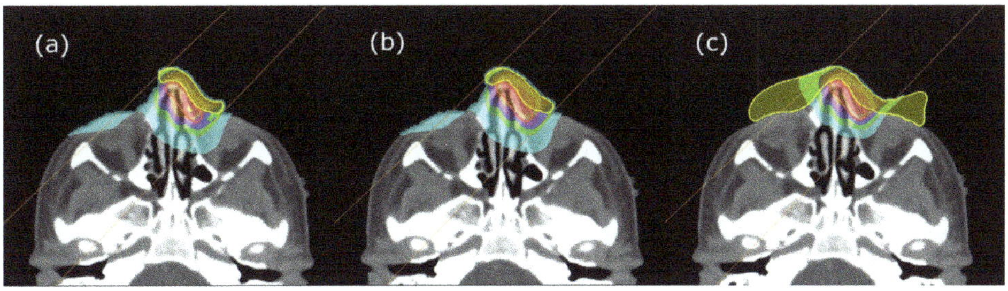

Figure 9.3. Electron therapy dose distributions calculated using Raystation treatment planning system, for a skin cancer treatment of the nose where bolus (yellow outline) has (a) uniform thickness, (b) variable thickness to achieve a uniform dose distribution despite density heterogeneities in underlying tissues and (c) variable thickness to achieve a compensator effect, maximising target coverage while minimising dose to underlying organs at risk and sparing the eyes.

a 12 MeV electron treatment of mycosis fungoides of the face, respectively. Su *et al* [60] achieved a 38% reduction in mean kidney dose and Zhao *et al* [32] achieved a 10%–20% reduction in mean dose to the eyes when a uniform bolus was replaced with patient-specific 3D-printed boluses. Several multi-patient dosimetry studies have also demonstrated how an optimised 3D-printed bolus can improve target coverage while sparing underlying structures for other anatomical sites [36, 55, 61].

The increased irregularity and thickness of these electron compensator boluses compared to photon boluses whose only purpose is negating the skin sparing effect (see figure 9.2) may lead to comparatively long print times. For example, Zhao *et al*'s [32] abdominal electron bolus took nearly 24 h to print and Lee *et al*'s [61] breast electron boluses took around 10 h each. To minimise this fabrication time, while also reducing the usage of 3D-printer filament, several authors have suggested the use of 3D printing to make shells for moulding electron bolus from silicone [36, 55] (also see chapter 3). The use of silicone is also expected to show improved flexibility, and therefore potentially achieve improved bolus fit compared to commonly used MEX filaments such as PLA and ABS [36, 55], although it should be understood that the flexibility of silicone, like the many flexible TPU-based filaments now available for MEX printing, varies substantially with bolus thickness and may in fact be negligible in the thicker parts of an electron compensator bolus.

Since an electron compensator bolus is specifically designed to shape the dose distribution, often accounting for variations in the density of underlying anatomy [32, 58], it is especially important to achieve a bolus density, geometry, and placement during treatment which matches the bolus modelled in the treatment plan. For this reason, it is common for studies involving 3D-printed electron bolus to include a quality assurance step where the patient is CT scanned with the bolus *in situ* for the purpose of checking fit and alignment in addition to print density and homogeneity [32, 60–62]. Using this new scan, the planned dose distribution can be recalculated and checked against the initial planned dose [55, 63], provided that additional checks of material properties and electron dose calculation accuracy are completed in advance [33], as described below.

While the use of pre-treatment cone-beam CT imaging as a geometric check of bolus fit is comparatively straightforward [32, 63], the use of a kV (planning) CT to perform dosimetric checks or verify electron density is less straightforward for electron radiation therapy than it is for photon radiation therapy. Whereas the kV photon interactions from the imaging beam and the MV photon interactions from the photon radiation therapy beam are, respectively, dominated by photoelectric and Compton interactions, electron radiation therapy interactions are mediated by the Coulomb force between the travelling electron and the atomic electrons in its path [64]. Electrons, as very light charged particles, are subject to multiple scattering [64]. The role of multiple Coulomb scattering in the delivery of radiation therapy dose from electron beams leads to concerns about the effects of 3D-printing material stopping power and possible air gaps between bolus and skin on the reliability of radiation therapy treatment planning dose calculations.

Before using kV CT to verify the material properties of an electron radiation therapy bolus, it is important to establish that the Hounsfield units (HU) reported by the CT image provide a sufficiently accurate representation of the effect that the particular bolus material has on the electron therapy beam. Generally, this is most easily achieved by using a 3D-printing material that has an electron density, mass density, and effective atomic number that approximate the electron density, mass density, and effective atomic number of water. For materials that differ from water in any of these parameters, it may be necessary to identify an appropriate HU override to be used for the bolus when planning electron treatments. For example, Burleson *et al* used stopping-power corrected parallel plate ionisation chamber measurements through various thicknesses of 3D-printed material to obtain electron beam tissue-maximum ratio (TMR) and percentage depth-dose profile (PDD) measurements, which were then compared with corresponding TMR and PDD calculations in their treatment planning system, to identify the HU override that caused the planning system to produce TMR and PDD profiles that matched the measurements [33]. Using Burleson *et al*'s [33] method, it should be possible to identify HU override values that will produce accurate electron radiation therapy dose calculations in the presence of different types of 3D-printed bolus.

The effects of air gaps under electron bolus have previously been examined for the large centimetre-scale gaps that can occur when a generic sheet bolus is placed on irregular surfaces [65], but further examination of these effects for the small millimetre-scale air gaps that are observed under patient-specific 3D-printed bolus is advisable. Note that it is not sufficient to recalculate the planned dose using the treatment planning system to test these effects, unless the accuracy of dose calculations provided by the treatment planning system have been specifically verified when small air gaps are present (see Kairn *et al* [39] for a demonstration relating to photon radiation therapy).

9.3.4 Proton radiation therapy

There are two obvious applications of 3D-printing for passive scattering proton treatment systems. These are the fabrication of range shifter wheels [66, 67], to achieve prescription coverage of the treatment target, and compensators [23, 24], to achieve beam shaping at the distal edge of the target and improve downstream organ-at-risk sparing. The use of 3D printing allows detailed patient-specific wheels and compensators to be produced without producing the material wastage [24] or requiring the physical space and technical skills [23] demanded by conventional milling systems.

For pencil-beam scanning systems, the key application of 3D printing is in the fabrication of pre-absorbers, to achieve prescription coverage of shallow targets despite the high energy (typically 70 to 250 MeV [21]) of many contemporary scanning proton beam systems. These pre-absorbers can be in the form of uniform-thickness range shifters, bolus, or variable-thickness compensators [25, 26].

When designing compensators, boluses, and range shifters that are intended for placement outside the proton treatment nozzle, it is important to consider the effect of the beam modifier position on the proton beam scatter characteristics and beam spot size. It is desirable to position the modifier as close as possible to the patient, preferably on the patient's skin rather than attaching the modifier to the nozzle or snout [25, 68, 69], to minimise beam spot degradation and consequent lateral penumbra broadening, while achieving a minimal increase in out-of-field scatter from the modifier [26]. This change in practice, illustrated in figure 9.4, creates an additional opportunity for 3D printing. For example, the use of a uniform-thickness bolus placed directly on the patient's skin produces substantially less focal spot degradation than a flat range shifter at the nozzle, while achieving the same pre-absorption effect on the beam [25]. Then, if the bolus can be 3D printed to fit the patient's surface contours, it becomes possible (and potentially desirable to help reduce spot numbers and beam energy switching when treating moving targets [22, 70]) to augment the bolus with variable-thickness compensation, analogous to the electron bolus demonstrated in figure 9.3(c).

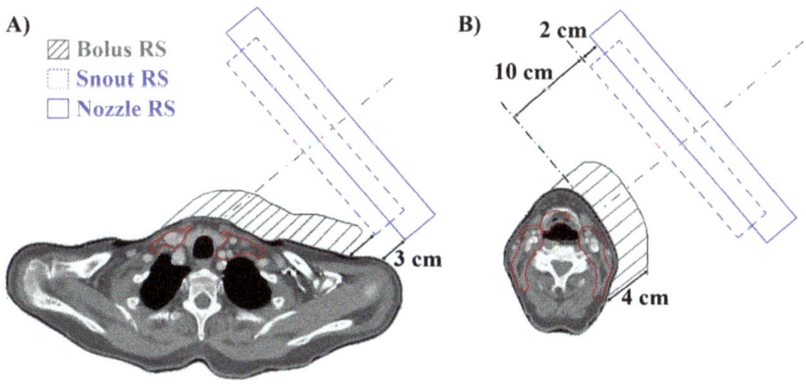

Figure 9.4. 'Range shifter (RS) applied as bolus and mounted on a snout or on the nozzle. (A) 3?mm air gap was maintained between the body contour and the bolus RS to include nonideal set-up in the modelling. A 3 cm safety gap was maintained for the snout and nozzle RS to avoid collision with the patient. (B) This translated into a maximum air gap of 10 cm and 12 cm, respectively, on the more cranially located slices.' [25]. Reprinted from [25], Copyright (2018), with permission from Elsevier.

For all of these applications, whether for use in passive or pencil-beam scanning systems, care must be taken to ensure that the 3D-printed materials either match the properties defined in the treatment plan (for compensators or boluses) or are able to accurately produce the planned properties of the proton beam (for range shifters and wheels). To verify this accuracy, recent studies have used dual energy CT to evaluate relative electron densities, effective atomic numbers, and proton stopping-power ratios of tested materials [71], in addition to measuring downstream PDD profiles [23, 24, 66, 72] and proton beam spot sizes [69, 71]. Methods such as these should be used to verify that the dosimetric effects of chosen 3D-printing material are correctly represented by the treatment planning system (also see section 9.3.3). It is also important to verify the dosimetric effects of the printing material across the expected range of proton beam energies, given that some materials produce substantially different effects on photon beams at different energies [73, 74] (see chapter 8).

Additionally, care must be taken to verify the fidelity of the 3D-print geometry (including fine-scale details [66]) with reference to the intended design. Depending on print density, it may also be necessary to carefully evaluate the effects of infill pattern and orientation, so that any desired range shifting effects are achieved consistently. Repeating PDD measurements with samples positioned at different orientations is advisable, especially after any changes in printing method. Further safety tests of 3D-printed materials may also be advisable, depending on the print material and application. For example, Michiels *et al* [71] performed mechanical testing on 3D-printed samples intended for use as proton patient immobilisation equipment and Lindsay *et al* [66] used gamma spectrometry to test for 3D-print activation after proton beam irradiation.

An emerging treatment technique is ultra-high dose rate FLASH radiation therapy [75]. Achieving the 40Gy s^{-1} dose rate required for FLASH radiation therapy is challenging for pencil-beam scanning proton accelerators, primarily due to the time delay when switching energy layers. A simple strategy for avoiding energy switching is to 'shoot-through' the target and treat entirely with the plateau region of a single high-energy layer, but this removes all dose conformality proximal and distal to the target [76]. Recently, patient-specific 3D-printed range modulators have been designed to overcome this shortcoming [77]. The printed filters feature a series of fine ridges and pins to shift the depth of the Bragg peak as the single-energy pencil-beam is swept across the field. Optimising the design and manufacturing of these filters is an active area of research [78].

With appropriate planning, careful implementation, and stringent testing (see chapters 7 and 8), 3D printing has the potential to enable efficient and clinically effective fabrication of proton treatment accessories, for all of these patient-specific applications.

9.4 Custom shielding

9.4.1 Clinical context

Patient-specific shielding and generic shielding devices are often used during radiation therapy treatments to block radiation from reaching sensitive anatomy or implanted electronic devices, or to collimate the radiation beam. Despite its known toxicity [79, 80], lead remains widely used for radiation shielding applications, even in applications involving patient contact. Other heavy metals, including cadmium—a known carcinogen [81]—are also used in some low melting point alloys in combination with lead when forming electron beam shaping apertures (see section 9.4.3).

Lead has excellent radiation shielding effects thanks to its high atomic number, density and electron density. These properties, in combination with lead's softness and malleability, have resulted in the widespread use of lead for radiation therapy patient shielding applications. However, the continued use of lead is associated with serious health risks that become less tolerable as more alternative shielding materials are identified [82–84].

The detrimental effects of lead exposure on the developing brains of children are well known, but in recent decades the effects of prolonged (e.g. occupational) exposure to lead in adults have been investigated via epidemiology, epigenetics, and occupational studies and longitudinal studies [79, 80, 85]. Lead preferentially sequesters in bones and then kidneys [79], leading to anaemia and kidney disfunction [79]. Adult and elderly cognitive function has been found to be inversely correlated with tibia lead levels [85], with exposed individuals showing increased cognitive impairment compared to non-exposed individuals, and with the degree of impairment increasing with age [85]. Lead may affect cognition by supplanting the role of calcium in ion transfer in the brain [79], while also potentially affecting gene expression in ways that contribute to the development of serious and terminal neurological diseases [85]. Lead exposure can also produce fertility effects, including reduced birth weight and possible postnatal developmental delay [80], as well as increased risk of miscarriage and reduced male fertility [79].

Lead and cadmium exposure can occur via inhalation, ingestion, and dermal absorption [79, 81]. Both metals accumulate in the body, so that frequent low-level exposure can become hazardous [80, 81]. Obviously, work that involves touching, hammering, or rasping (filing) radiation therapy shielding devices containing lead or cadmium should be minimised as much as possible and should only be carried out with the use of appropriate personal protective equipment. Furthermore, radiation therapy workers who are routinely involved in fabrication of these devices should receive regular blood lead level tests [80]. Ideally, the use of these materials for radiation therapy shielding should be phased out as non-lead and non-cadmium alternatives become available.

Obviously, radiation therapy patients should be protected as much as possible from direct contact with shielding devices containing lead or cadmium. In addition to safety, consideration should also be given to patient comfort when it comes to the fabrication and use of radiation therapy shielding devices.

For kV apertures in particular, casts of the patient's anatomy are often needed for the formation of appropriately shaped cutouts. The creation of these casts (usually in plaster or thermoplastic) involves pressing wet, warm, or otherwise uncomfortable materials onto areas of the patient's skin that can be diseased, sensitive, and/or painful.

The use of 3D printing in this area has the potential to reduce staff contact with lead and other toxic metals during the fabrication and use of shielding blocks and apertures, while also, if combined with surface scanning, reducing the need for direct casting of patient positive models.

9.4.2 Shields

Patient-specific shielding blocks and generic shielding devices are most often used for kV photon therapy and MeV electron therapy, and rarely for MV photon or MeV proton therapy. This is because the weight of lead required for shielding of kV photon and MeV electron beams is generally tolerable by the patient when the block is placed directly on their skin. Meanwhile, the thickness of lead required to exceed the buildup thickness and provide effective shielding of MV photon beams means that MV photon shielding blocks are generally used only for out-of-field applications where either (a) any shielding may be more worthwhile than none, such as shielding radiation sensitive implanted electronic devices [86], (b) the shielding weight can be externally supported, such as by a testicular shield stand [87], or (c) the radiation protection need is sufficient to justify suspending large volumes of lead over the patient's body, such as in some rare cases where radiation therapy treatments of pregnant patients are required [88].

The lead thicknesses required for shielding lower electron beam energies are in the millimetres—electron beams with nominal energies less than 10 MeV can be shielded using lead thicknesses less than 5 mm to give transmissions of less than 5%. (A rule of thumb for electron shielding is that the required lead thickness in millimetres to reach the electron practical range is approximately 0.5 times the nominal electron beam energy in MeV [64, 89].) For these lower energy electron beams, there are evidently many opportunities to replace the use of lead shields with custom 3D-printed shields made from more benign materials, such as PLA or other plastics.

For example, total skin electron therapy (TSET, also known as TSEI or TSEBT) treatments are delivered to standing patients using nominal 6 MeV electron beams with a 4 MeV expected energy at the patient's surface [90], and require shielding of several anatomical areas to prevent side effects or mitigate hotspots [91]. Rahimy *et al* [92] noted that the 2 mm thickness of lead that was being used to shield the scalp in their TSET treatments could be replaced with a 23 mm thickness of water-equivalent material or 20 mm thickness of (slightly denser than water) PLA. Accordingly, they used 3D optical surface scanning (see section 9.4.4 and chapter 4) to design patient-specific scalp shields that were 3D printed using PLA [92]. Rahimy *et al* [92] found that the 3D-printed shields had a clinically useful adaptability compared to their generic lead 'helmet', allowing lesions at the nape of the neck to be treated while sparing hair follicles and avoiding alopecia. The main limitation of the optical scanning method, scan inaccuracies produced by hair (see section 9.4.4), was avoided by Elmahmoud *et al* [90] who used a CT scan of their patient's head to design a bolus that was subsequently shown to fit the patient 'snugly' and was comfortably worn throughout a TSET treatment that ultimately produced a near-complete response with no alopecia at 6 months. Clearly, opportunities exist for 3D printing non-lead shielding for other parts of the body that are frequently shielded during TSET, such as fingers and scrotum [91].

There are also opportunities to use 3D printing to replace the use of lead eye shields used in electron radiation therapy of the face, especially if 3D printing is already being used to create patient-specific bolus for these treatments (see section 9.3.3). For example, Sasaki *et al* [53] and Chiu *et al* [55] have, respectively, used 3D printing to produce bolus for 9 and 18 MeV electron treatments of the face, with gaps or recesses designed to accommodate lead eye shields. While fabrication of a usable non-lead shield for electron beam energies greater than 10 MeV may be challenging, it should be possible to use dual-nozzle MEX printers to fabricate combined bolus and shielding devices that achieve the required degree of shielding using non-lead shielding materials, such as PLA-tungsten composites (see chapter 3). Careful design of such devices could also allow the weight of the shielding to be distributed across the bolus, while placing some bolus material between the shielding and the patient, to help absorb unwanted low-energy electron scatter or bremsstrahlung photons.

As noted earlier, shielding devices are rarely used during MV photon and proton treatments, due to the increased thickness and therefore weight of shielding required to shield the beams. Nonetheless, one possible application of 3D printing in this area is the fabrication of plastic replicas of MV photon and proton shields for sensitive anatomy or implanted electronic devices. These replicas are useful during the initial planning CT session to evaluate patient comfort (relating to the shape but not the weight of the intended shield) and to provide an accurate representation of the shield in the planning CT, which can be contoured and overridden to improve dose calculation accuracy, while avoiding the substantial reconstruction artefacts that would result from the use of a high-density shield during CT scanning [93, 94]. Burke *et al* [94] have also demonstrated how the use of a 3D-printed replica of a generic testicular shield can be used instead of the heavy lead shield to minimise testicular dose by simply shifting the scrotum out of the path of the radiation beams. (Use of 3D printing for patient positioning and immobilisation is discussed in section 9.5).

As always, careful assessment and ongoing quality control of the dosimetric effects of the 3D printed materials is necessary to demonstrate their safety and efficacy (see chapter 8). For 3D printed shields, it is particularly important to investigate the production of contaminant radiation from the shields, identify density overrides necessary to achieve accurate treatment planning dose calculations, and frequently confirm the reduction in beam transmission achieved by the shields. It is also necessary to maintain clear records and strict administrative controls regarding which type of shield is suitable for use in which beam because misuse (e.g. a 6 MeV electron shield in a 6 MV photon beam) can result in dose enhancement rather than radiation shielding.

9.4.3 Apertures

A further purpose of shielding in kV photon and MeV electron radiation therapy is in the field shaping of the treatment beam, using shielded apertures. kV photon

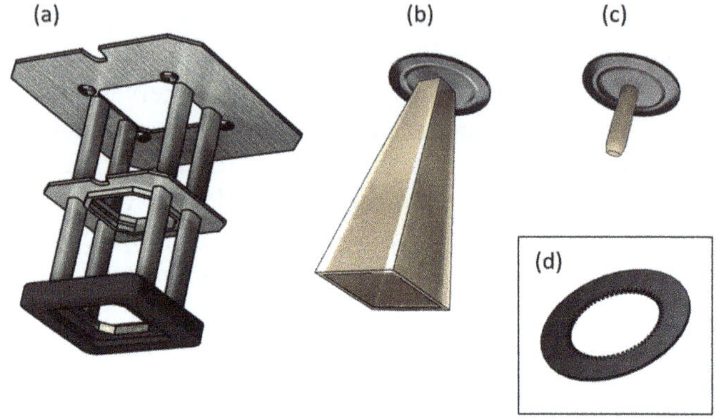

Figure 9.5. Diagrams showing (a) electron applicator containing a patient-specific cutout, (b) orthovoltage photon applicator, (c) superficial photon applicator, and (d) example kV (superficial or orthovoltage) patient-specific cutout, not to scale.

radiation therapy beams and MeV electron radiation therapy beams are highly susceptible to scattering in air, and so are most often delivered using a combination of applicators, which collimate the beams after they respectively exit the x-ray tube or linac head, and patient-specific apertures or 'cutouts' (also called inserts, blocks, or masks) that shape the beams at or close to the patient's surface. Figure 9.5 shows diagrams of applicators and cutouts for kV photon radiation therapy and electron beam radiation therapy.

For superficial photon beams (with energies in the range 50–100 kVp) and orthovoltage photon beams (with energies in the range 100–500 kVp) cutouts are placed directly on the patient's skin after being manually shaped over models (positives, see section 9.4.4) of the relevant anatomy. These kV cutouts are most commonly formed using lead sheets with thicknesses of 0.2–3.5 mm, depending on beam energy and penetration. Although paint, tape, or plastic wrap are frequently used to protect patients from contact with the lead, the need for staff to handle, shape, hammer, cut, and file the lead sheets remains.

To avoid the hazards arising from making kV cutouts using lead, Sharma *et al* [95] have proposed using specialised 3D printers to create cutouts from other metals, with an acknowledgement that such direct metal printing systems may currently be prohibitively expensive for radiation therapy applications (see chapter 2). More recently, Crowe *et al* [96] have demonstrated the potential for PLA-composite prints containing tungsten and copper to be used as shielding for superficial and orthovoltage beams (in the case of the PLA-tungsten composite) or superficial beams only (in the case of the PLA-copper composite). Obviously, the reduced density of the PLA-composite filaments required the shielding thicknesses to be increased when compared to lead [96]. To achieve a sufficient shielding effect for clinical use (i.e. 5% transmission), the thickness of the printed PLA-tungsten material needed to be increased to 1.6–3.0 times the thickness of lead used for the corresponding kV beams and the thickness of the PLA-copper composite needed to be increased much more [96].

For MV electron beams (which usually have nominal energies in the range 4–20 MeV) cutouts are attached to the end of the electron applicator (see figure 9.5) and are conventionally created using heavy metal alloys containing substantial proportions of lead. Low melting point alloys (LMPA) used for this purpose include Cerrobend (50% Bi, 26.7% Pb, 13.3% Sn, and 10% Cd) and its cadmium-free alternatives (e.g. 52% Bi, 32% Pb, 16% Sn or 50% Bi, 31% Pb, and 19% Sn). The standard process of fabricating LMPA electron cutouts consists of designing the cutout to achieve a specific treatment goal in a radiation therapy treatment planning system, and then exporting the aperture shape which is cut from styrofoam (or similar material), and placed in a cutout frame and used as a mould for the pouring of molten LMPA. Once set, the styrofoam is removed and the remaining LMPA in its cutout frame becomes the cutout for the electron radiation therapy treatment. Given the several serious hazards involved in the standard LMPA process, it is unsurprising that fabrication of electron cutouts has been identified as a major application for 3D printing in radiation therapy. Albantow and Brown's [28] 2022 survey of the 112 radiation therapy centres in Australia and New Zealand found that 28.6% of centres were using 3D-printed electron cutouts supplied by third parties and one centre reported the use of a 3D-printing system for in-house electron cutout production.

Methods for using 3D printing to minimise the hazards and increase the accuracy of electron cutout fabrication range from fabricating plastic (PLA or ABS) jigs for improving the precision and reproducibility of the LMPA moulding process [97, 98] (see figure 9.6(a)) to avoiding the use of LMPA altogether through the use of

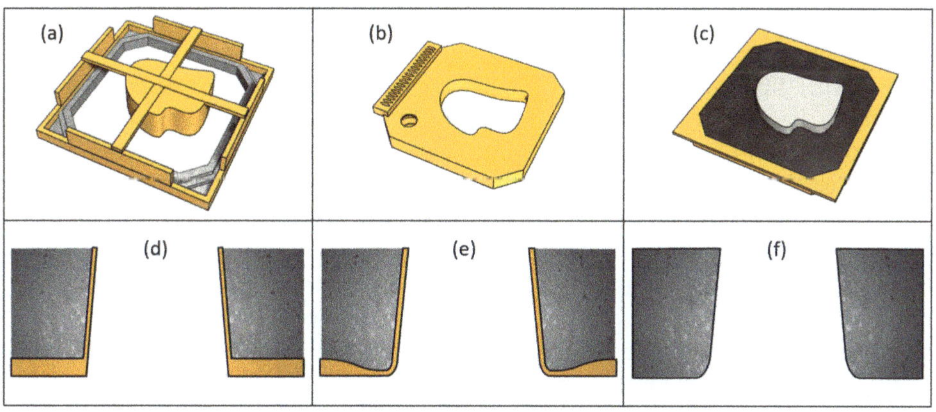

Figure 9.6. Alternative designs for 3D-printed electron cutouts: (a) 3D-printed frame with precise aperture positioning control for improved LMPA pouring [97, 98], (b) 3D-printed housing into which tungsten ball bearings can be poured [99, 100], (c) 3D-printed frame and dissolvable PVC aperture mould, for setting of tungsten carbide powder in thermoset plastic [89]; and alternative methods for handling aperture edges in 3D-printed electron cutouts: (d) straight plastic edges and thick base, for supporting tungsten ball bearings [99], (e) plastic edges with thinner base around the aperture to improve out-of-field dose profile [99], (f) no plastic walls or base, due to the use of dissolving PVC and fixed tungsten carbide [89].

tungsten ball bearings or powders in 3D-printed frames [89, 99, 100] (see figure 9.6(b)). 3D-printed electron cutouts that replace LMPA with tungsten as the shielding component successfully avoid the use of molten metals, lead alloys, and cadmium in the cutout production, and avoid the requirement for staff to handle lead alloys when setting up election treatments for patients. However, the use of a granular shielding material (1.5–2mm tungsten alloy ball bearings [99, 100] or tungsten carbide powder [89]) requires careful investigation of density consistency and consideration of the handling of aperture edges to ensure that the cutouts achieve their intended dosimetric effects [89, 99, 100].

For example, Skinner *et al* [99] showed how using a 3D-printed plastic frame with a 1.5 mm thick base could produce a localised increase in out-of-field dose beneath their tungsten ball bearing cutout, which could then be mitigated by rounding the cutout edge at the lower end of the aperture and making the base thinner in in this region (see figures 9.6(d) and (e)). Later, Breitkreutz *et al* [100] refined the loose ball bearing design by adding a solid lid to confine the tungsten ball bearings, as well as a brass or copper frame around the outer edges of the cutout, to mitigate possible ball bearing redistributions due to gravity, with gantry rotation. Skinner *et al* [99] and Breitkreutz *et al* [100] both highlighted the need for measurements of cutout weight as part of the cutout quality control processes to make sure that the shielding densities were consistent with the calculations. Both groups also recommended reducing the thickness of the 3D-printed aperture walls to minimise electron scatter from the plastic. Skinner *et al* [99] also noted the need to match the angulation of the aperture walls with the angle of beam divergence.

While 3D-printed aperture walls are needed to confine loose tungsten alloy ball bearings, the need for these walls as well as the 3D-printed base and lid of the cutout frame can be eliminated by fixing the tungsten in place. For example, Leat *et al* [89] have demonstrated a process where 3D printing was used only to fabricate a PLA frame (to interface with the electron applicator) and a polyvinyl alcohol (PVA) temporary casting core (similar to the conventional styrofoam cores used in LMPA moulding), which were used together as a mould into which tungsten carbide powder was compacted and fixed using a thermoset polymer. After polymer curing, the PVA core was dissolved leaving the completed cutout with no plastic wall lining the inner surface of the aperture [89], as illustrated in figures 9.6(c) and (f).

When developing or using non-standard (non-lead) electron and kV photon cutouts, it is important to assess the thickness of material required to achieve the intended shielding effect, as well as the effects on low-energy electron scatter or the production and attenuation of bremsstrahlung. Meanwhile, kV photon cutout options should be benchmarked against lead, and non-LMPA (e.g. tungsten) electron cutouts should be benchmarked against LMPA, with consideration for the effects of the different materials on in-field dose as well as out-of-field transmission and penumbra shapes. These assessments should be performed using all clinically relevant beam energies. Finally, if shielded apertures are constructed using ball bearings or powders, it is important that frequent systematic quality control tests of shielding density, consistency, and homogeneity are completed and documented.

9.4.4 Positives

For kV radiation therapy in particular, where the direct 3D printing of shielding materials is possible but still under investigation [96], the standard clinical practice for patient-specific shielded aperture fabrication remains as follows. The patient attends a cutout fabrication appointment, where a cast is made of the surface to be treated and the surrounding area using plaster of paris, or thermoplastic, or another casting material [95, 101, 102]. The patient must remain still, often lying flat (depending on treated anatomy), while the casting material is gently pressed onto their skin and while waiting for the cast to set. The face is a common site of skin cancer, which can require one or both eyes to be covered by the cast and may require special handling of nostrils and/or mouth, so that the patient can continue to breathe comfortably during casting. After the casting process, which may take more than one attempt and may take up to an hour, especially if interrupted or restarted due to patient stress or discomfort, the patient is cleaned and released. The remaining steps require staff time and manual handling of hazardous lead, but are completed in the absence of the patient: the cast is used to create a positive mould of the patient's anatomy, usually using plaster of paris, which is allowed to set and then used as a last for a sheet 0.2–3.5mm thick sheet of lead to be pressed and beaten into the shape of the patient's external contour before the planned treatment aperture is cut from the lead. The localisation of this aperture is also challenging, given the need to reliably transfer markings on the patients' skin to the lead sheet, often achieved using a flat acetate sheet and without the use of a planning target volume to account for geometric uncertainties introduced in this process [103]. Clearly, this process involves substantial aperture placement uncertainty, substantial discomfort for the patient, and a substantial time commitment from staff.

Several groups have investigated the use of 3D printing in combination with optical surface scanning to dramatically reduce the patient discomfort and staff time involved in the fabrication of patient-specific lead cutouts for kV radiation therapy, while simplifying the aperture placement process [95, 101, 102]. These studies have used handheld optical scanning devices that contained cameras paired with structured light sources (including infra-red [95, 101] and laser [102]), which were used to reconstruct 3D surface images [95, 102] (also see chapter 4).

Crowe *et al* [104] have shown that such images can be obtained and reconstructed within a few minutes per anatomical site, while Sharma *et al* [95] have used a metal hoop as a simple gantry for their optical scanner, allowing scans to be completed in around a minute. Similarly, Briggs *et al* [102] have completed optical scanning of a patient's face within a five minute appointment, including time taken to verify the scan. In all of these studies, detailed 3D maps of skin surfaces were acquired without any need for touching of the patients or study participants, other than some gentle marking of planned aperture positions [95, 101, 102, 104].

Crowe *et al* [104] have demonstrated how positions of markings drawn on the skin can be transferred to the 3D surface maps produced by optical scanning, via either the photographic images (referred to as 'textures') that overlie the surface topography, or by placing 3D markers (e.g. wires) on the skin so that they are

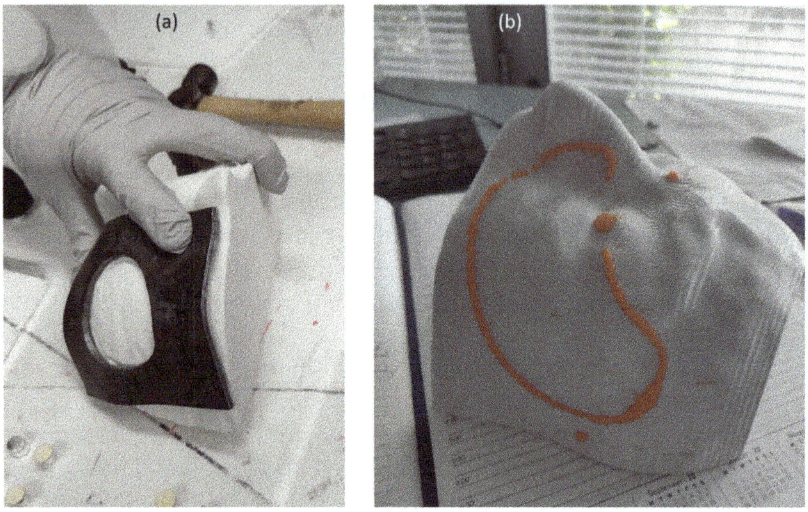

Figure 9.7. Photographs of 3D-printed positives showing (a) use in shaping a lead cutout, and (b) planned aperture position marked using a contrasting filament from dual-nozzle MEX printer.

automatically included in the topography itself. In a case study by Briggs *et al* [102] and a 10 patient study by Sharma *et al* [95], the optically scanned 3D surface maps were modified to include indentations showing the intended aperture borders and converted into STL files using open-source software (Blender [102] and Meshmixer [95, 104]) and then 3D printed to produce the final positive models for lead shielding fabrication. Similar optical scanning methods have also been used to create 3D-printed positive models for use in fabricating shielding for electron radiation therapy treatments [53, 92]. Examples of positive models are shown in figure 9.7.

Print times for positive models of patient faces have been reported ranging from 12 h [95, 101] to 30 h [102], although it should be understood that the print process requires no attention from staff other than checking the correct deposition of the initial layers and may be completed overnight, provided that safe operation of the 3D printers (e.g. functional thermal runaway protection) has been established (see chapters 6 and 8).

Geometric accuracy of 3D-printed patient positives is very important if they are to replace plaster moulds in the production of patient-specific lead shielding. When these 3D-printed positives rely on optical scanning for their designs, it is imperative that the reliability of the optical scanning systems is carefully evaluated, such as by scanning simple objects with known geometries and different colours, under different ambient light conditions [104]. Sharma *et al* [95] have also described their process of optically scanning phantoms with simple and complex surfaces, and comparing the resulting surface maps against high-resolution CT scans of the same phantoms. It should also be understood that results of optical scanning may be affected by reflections from damp, broken, or bleeding skin, depending on the light source used. Additionally, issues arising from the scanning light path being blocked by thick hair [53, 92] or clothing [104] are also to be expected.

A programme of routine quality control testing of the resulting 3D-printed positives should also be created and documented. This may include CT scanning each positive and comparing the surface against the STL file from which it was printed [53] or simply creating the shielding device and having the patient or treatment staff evaluate its fit, comfort, and reproducibility [102].

9.5 Immobilisation

9.5.1 Clinical context

The importance of immobilisation equipment to maximise the precision, safety, and effectiveness of external beam radiation therapy has been widely recognised for decades [105]. To be safely treated using radiation exposures that may last minutes or hours, each patient must remain stationary in the same position that they were in when their treatment planning images were acquired. Immobilisation equipment is used to set up each patient reliably in their planned position at each radiation therapy fraction, and to keep them still throughout the delivery of every treatment fraction. Immobilisation equipment needs to be as comfortable as possible, while affecting the radiation beam as minimally (or predictably) as possible. Immobilisation during treatment must either use the exact same devices that were used during the acquisition of images for treatment planning, or must match the geometry, electron density, and degree of flexibility of the devices that were used during the acquisition of images for treatment planning.

Immobilisation equipment ranges in complexity from simple knee rests and pillows to complex positioning devices and mask systems. Immobilisation equipment can be used externally or internally, can be fabricated to match a single patient, or be designed for use by multiple patients receiving a particular type of treatment, and can be intended for single use or for repeated uses with one patient or many. Consequently the patient safety considerations for different types of immobilisation devices vary (see section 9.2), and it is also important to consider the mechanical safety of any 3D-printed devices that are intended for immobilisation purposes.

9.5.2 Mechanical safety

For support and immobilisation applications, it is important to consider the mechanical properties of 3D-print designs and the possible effects of fabrication techniques (including selection and orientation of infill patterns), in concert with questions of print material and density that are encountered for other 3D-printing applications (e.g. see chapter 10).

Previous studies have demonstrated that the strength of mechanical parts fabricated via MEX printing can be reduced by more than 30% when compared to the same mechanical parts fabricated by directly moulding the same materials [106, 107]. The strength, elasticity, and other mechanical properties of 3D prints can also be affected by infill density, infill pattern/orientation, layer thickness, and material [106–109]. Mechanical testing results reported by Tankella *et al* have suggested that 3D-prints fabricated using polycarbonate are comparatively strong, while printed ninjaflex samples can stretch by 800% without breaking, and nylon prints can be stronger than both ninjaflex TPU and semiflex TPU while also being more flexible than ABS, HIPS, T-Glase, and polycarbonate [106]. Filament colour has also been observed to have small but measurable effects on mechanical testing results, due to the effects of microscopic colouring agents on the crystallisation of polymers in the filaments [106, 110].

Several studies have also used computational engineering tools such as numerical (finite element) modelling, to investigate the effects of print design on strength and robustness [109, 111, 112]. For example, Stava *et al* [112] used computational modelling of loads produced by a pinch-grip, a worst-case example of the load on a small object due to manual handling, to produce a system for automatically adding thickness, inserting struts or creating voids within 3D-print designs, to minimise likelihood of breakage. Lu *et al* investigated the potential for using complex internal voids (voronoi patterns) based on internal stress maps from computer modelling to produce built-to-last objects with optimised strength-to-weight ratios [111] (e.g. see in figure 9.8). Clearly in the context of radiation therapy support and immobilisation equipment, where low-densities are often desirable to maximise radio-transparency, these techniques for maximising robustness have the potential to be especially useful.

If the immobilisation device is intended for extended or repeated exposure to the radiation beam, then additional testing for possible effects of radiation on samples of the 3D-printed material is also advisable. For example, Van der Walt *et al* [42] and Kairn *et al* [43] both performed simple tests of geometry and density before and after extended irradiations of 3D-printed materials, and Michiels *et al* took the important extra step (critical for immobilisation devices) of assessing mechanical properties before and after radiation exposure [71].

Finally, all methods and results of mechanical testing (as well as all other safety testing, see section 9.2) of 3D-printed samples or devices intended for immobilisation use should be documented, and kept as a record of quality management activity (see chapter 7) and as a baseline for future mechanical testing work.

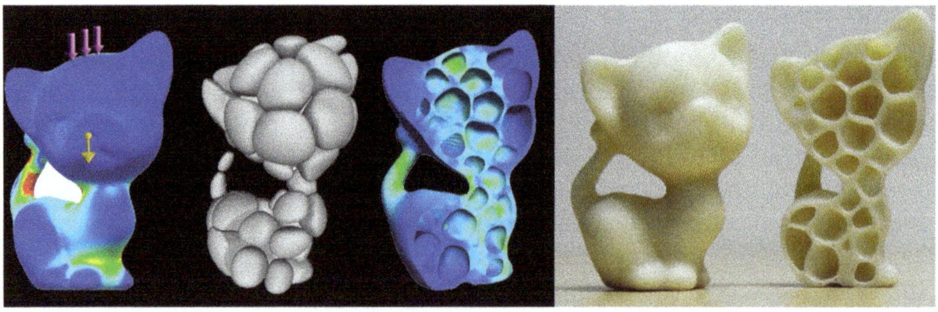

Figure 9.8. 'We reduce the material of a 3D kitten (left), by carving porous in the solid (mid-left), to yield a honeycomb-like interior structure which provides an optimal strength-to-weight ratio, and relieves the overall stress illustrated on a cross-section (mid-right). The 3D-printed hollowed solid is built-to-last using our interior structure (right).' Reprinted from [111], Copyright (2014) with permission from ACM (Association forComputing Machinery).

9.5.3 Patient supports

A wide range of commercial immobilisation systems are available for use in supporting and stabilising patients during external beam radiation therapy. While generic, these systems often also have movable and removable parts that allow devices to be arranged to fit the anatomy and movement restrictions of individual patients. For example, breast boards that are used to immobilise supine patients with their arms above their head often feature arm rests or hand holds that can be repositioned to make the patient as comfortable as possible.

For such adaptable systems, it may be possible to 3D-print custom attachments to suit particular clinical needs. For example, Baltz *et al* [113] developed a method for 3D-printing custom headrests that fit into the frameless adaptor on a GammaKnife machine to replace the use of thermoplastic for this geometrically challenging application. Others have developed methods for designing and verifying the geometric and radiological properties of 'headrest spacers' and wedges that fit into existing head-and-neck supports to achieve patient positioning that would not otherwise achievable with commercially available attachments [28]. While in-house design and fabrication of these accessories may be economical and time-efficient, care should be taken to ensure that appropriate legal reviews are completed and warranties are not voided.

Standalone 3D-printed support devices have been recently described for the immobilisation of extremities. For example, Albantow and Brown [28] noted that 'hand moulds' were among the most common applications of 3D printing in the radiation therapy immobilisation area in their 2022 survey of Australian and New Zealand radiation therapy centres, and Kairn *et al* [50] reported that a patient-specific 3D-printed foot support effectively immobilised the lower leg, entirely preventing rotations, while treating an extensive skin cancer. An example of this type of patient-specific foot support is shown in figure 9.9.

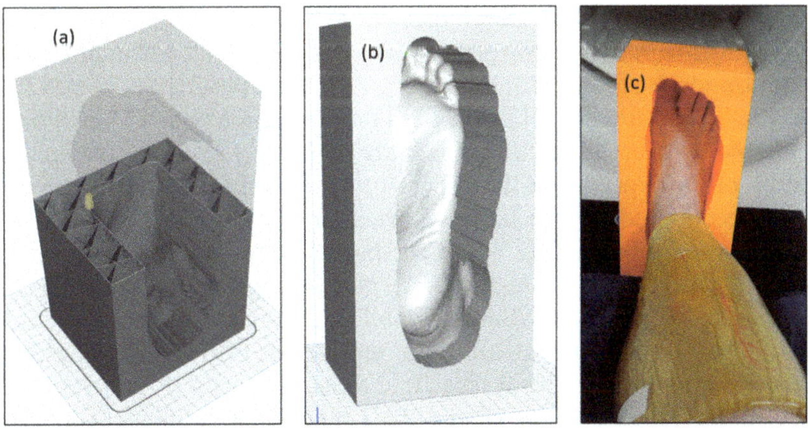

Figure 9.9. Patient-specific foot support: (a) and (b) screenshots of 3D-print preparation steps; and (c) foot support in use, in combination with a standard vacuum immobilisation bag.

When designing or evaluating 3D prints for potential use as patient supports, either in combination with commercial immobilisation systems or as standalone devices, it is important to assess geometric accuracy, compatibility with other immobilisation devices and patient comfort. Additionally, it is important to consider the effect that each device may have on the treatment dose, even if the devices are intended to be located outside the primary radiation beam. Hand moulds may be printed at unit density to provide scatter during electron radiation therapy treatments [28], whereas foot supports may be fabricated with low infill percentages to minimise scatter [50] and headrests are likely designed to provide minimal attenuation [113]. Assessments of print density and measurements of attenuation or backscatter may be needed to ensure that 3D-printed immobilisation devices are suitable for clinical use.

9.5.4 Immobilisation masks

Although immobilisation masks or shells generally moulded directly on the patient using thermoplastics have been reported in use for immobilising the thorax, abdomen, pelvis, and extremities [114–117], the most common use of these masks is in immobilising the head, neck, and often shoulders during head-and-neck radiation therapy [118]. Contemporary head-and-neck radiation therapy involves the delivery of complex and highly conformal radiation beams to treat targets that are surrounded by radiosensitive critical structures. Immobilisation masks are especially important to ensure the safety and accuracy of treatment delivery in these conditions.

However, the process of moulding the thermoplastic mask directly on each patient's skin has been highlighted as time-consuming for staff and uncomfortable or distressing for patients [119–122]. Miron *et al* [119] have pointed out that the forming, cooling (setting), and scanning of three-part masks used for immobilisation of some cranial stereotactic radiosurgery patients can involve the patient lying with their movement restricted and their mouth, eyes, and nose covered in thermoplastic mesh for up to an hour. The elimination of this difficult step from the radiation therapy treatment preparation process is a stated goal of many 3D-printed immobilisation mask studies [40].

While the literature on 3D-printed immobilisation masks includes several phantom-based studies and materials analyses [108, 123–125], the most reliable confirmation of the expectation that 3D printing can improve the patient experience comes from recent studies with human participants [119–121, 126].

Imaging studies with healthy volunteer participants have been used to develop workflows and evaluate inter-fraction positioning uncertainties [119, 120]. Deliberate attempts to move within immobilisation masks have provided information on the extent of possible intrafraction motion for non-compliant patients [119] and also measurements of the forces that these attempts to move may produce within the masks [126]. Mattke *et al*'s [121] reported treatment of a series of six whole brain radiation therapy patients using 3D-printed immobilisation masks demonstrated 'reliable and reproducible inter-fractional positioning accuracy', which compared favourably to the results from two control groups of patients who were treated with conventional thermoplastic masks. Most importantly, all of these studies demonstrated that patient-specific immobilisation masks could be designed and printed to fit each patient based on CT or MRI images, without the patient needing to be present for a moulding or fitting session.

An alternative method for using 3D printing to avoid the stress and discomfort of the immobilisation mask moulding session may be found in the process demonstrated by Pham *et al* [127], which is similar to the lead shield fabrication process using 3D-printed positives that was described in section 9.4.4. CT scans of 11 brain tumour patients were used to 3D-print replicas of the patients' heads and headrests. Thermoplastic masks were then moulded on these replicas rather than being moulded on the patients [127]. This method may be an ideal interim solution for radiation therapy centres who are moving towards 3D-printing immobilisation masks, or for centres wanting to avoid patient discomfort at the mask creation stage while also minimising changes to existing clinical workflows.

9.5.5 Displacement stabilisation

In addition to supporting patent anatomy from below (section 9.5.3) and stabilising patient anatomy from above (section 9.5.4), a third use of radiation therapy immobilisation devices is to displace and stabilise patient's anatomy from around or inside. Generally, the purpose of this displacement is to separate the radiation therapy target from the surrounding healthy anatomy, and thereby minimise irradiation of organs at risk. Examples of the use of 3D printing to achieve stable tissue displacement include the fabrication of holders and cups, to separate treated breasts from heart, lungs, and contralateral breasts [128, 129], or the printing of detailed patient-specific supports to keep cancerous fingers away from healthy fingers during radiation therapy [57]. An example of a 3D-printed immobilisation device that achieves finger displacement, for the treatment of a lesion affecting only one digit is shown in figure 9.10.

3D-printed immobilisation devices have potential applications in achieving stable displacement of internal anatomy. For example the use of a vaginal dilator to achieve sparing of the anterior vaginal wall as well as the vulva and urethra has long been recommended for radiation therapy of anal and rectal cancers [130, 131]. 3D printing could be used to produce bespoke dilators for this purpose given that 3D printing has been used to produce dilators for use in treating vaginal radiation injuries [132].

Recently, substantial research and clinical interest has been focused on using 3D printing for fabrication of intra-oral stents (including mouthpieces and bite-blocks) used during head-and-neck radiation therapy [4, 46, 133–137]. The oral cavity is both a challenge and an opportunity for investigators of 3D-printing applications in radiation therapy due to the radiosensitivity of tissues on all sides of the cavity, as well as the potential to effectively spare these tissues by displacing the mandible,

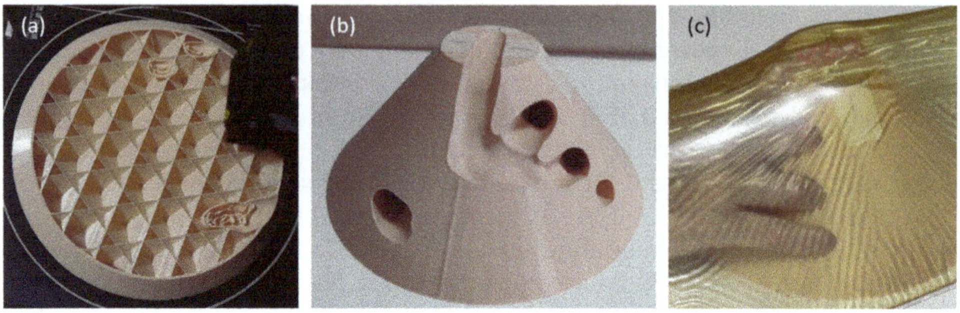

Figure 9.10. Photographs of a 3D-printed finger separation and immobilisation support: (a) during the 3D-printing process, showing the low-density internal structure, low position of the ends of the non-involved fingers and support structures being printed inside the finger holes; (b) completed 3D print, showing external support for separated finger to be treated; and (c) treatment setup, using conventional sheet bolus.

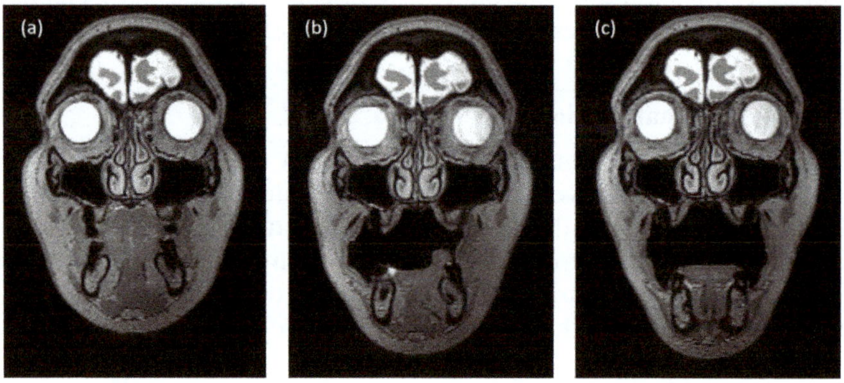

Figure 9.11. Coronal images of an MRI study participant [136] showing: (a) lack of separation between palate, tongue and buccal mucosa when a 3D-printed oral stent is not present, and separation between palate, tongue, and buccal mucosa when (a) one-sided and (b) two sided 3D-printed oral stents are present.

tongue, lips, or cheeks if such displacement can be achieved reproducibly and comfortably [138–140]. Figure 9.11 shows an example of how a 3D-printed intra-oral stent can spare the palate when treating the tongue or oropharynx, spare the tongue when treating the palate or nasopharynx, and spare the buccal mucosa when treating any central targets.

The use of 3D printing to fabricate intra-oral stents can have diverse objectives, from creating new generic devices for achieving a particular displacement effect [135], demonstrating customisable systems with parts that can be added, removed, or adjusted to suit each patient's anatomy or treatment objective [134, 137], to replacing the use of unstable (wax) or expensive (dental putty) materials in the creation of patient-specific devices [4, 133]. 3D printing may also have a role in improving patient comfort, given that study participants using 3D-printed oral stents have reported increased comfort and stability compared to wax [46, 137], provided that the 3D prints are not excessively rigid or slippery [136].

9.6 Brachytherapy

9.6.1 Clinical context

As mentioned in chapter 1, brachytherapy is a type of radiation therapy where a radioactive isotope is placed within or adjacent to the target volume. This is achieved using equipment such as catheters or applicators implanted or inserted into the patient or placed upon their skin. Brachytherapy treatment equipment is commercially available in a wide variety of generic forms, from cylindrical rods and tubes with interchangeable parts, to purpose-designed superficial applicators, and to rectilinear templates that allow parallel needles and catheters to be positioned at regular intervals throughout the target volume. While this equipment has been designed to maximise treatment adaptability and allow optimal dose distributions to be delivered accurately to suitable targets, improved brachytherapy treatments are achievable for non-standard targets by moving away from generic solutions towards more patient-specific applicator and template designs [141, 142]. In particular, the irregularities of post-surgery anatomy, uneven surfaces, and targets that are not optimally covered using parallel catheters may receive improved treatments using patient-specific brachytherapy solutions.

3D printing is an emerging technology in brachytherapy, with recognised potential to achieve improved conformity with patient anatomy, minimise air gaps and allow improved channel or catheter positioning, so that the radioactive source can be positioned optimally, and a delivered dose can more closely match the prescription and the treatment plan. To date, reported applications of 3D-printing in brachytherapy have included superficial [143–146], vaginal [5, 147, 148], oral cavity [6], and intraoperative head and neck [149] treatments. 3D-printed templates for interstitial catheter or needle insertion have also been customised to specific patient anatomy for sites such as vagina and cervix [150, 151], prostate [152], head and neck [142, 153–157], abdomen [158], and breast [159, 160].

The major obstacle to the widespread clinical use of 3D-printed brachytherapy applicators is the challenge of achieving a degree of sterilisation or cleaning that is sufficient to meet legislated standards and protect patients from contamination and infection risks. (Further information on these issues is provided in section 9.2.)

9.6.2 Superficial applicators

Commercially available applicators for brachytherapy treatments of the skin and superficial tissues take two forms, flap-based systems that are well-suited to delivering uniform dose to comparatively large and smooth surfaces via parallel catheters [161] and cone-shaped shielded applicators that are designed to treat small, approximately circular targets on flat surfaces [161]. Bespoke patient-specific surface moulds are needed to treat irregular lesions on curved surfaces in close proximity to sensitive organs [143, 161, 162] for which clinically acceptable dose distributions cannot be achieved using commercial applicators.

These patient-specific superficial moulds are conventionally fabricated using wax or thermoplastics with embedded catheters [143, 161, 162], but recent studies have

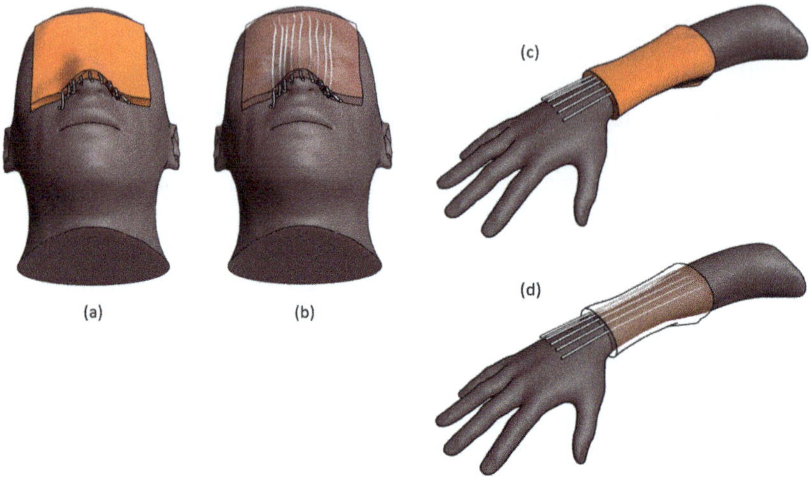

Figure 9.12. Illustrations showing designs for 3D-printed superficial applicators for treatments of skin cancers of the face (a) and (b), and forearm (c) and (c), with illustrations (a) and (c) indicating the external appearance of the completed prints and illustrations (b) and (d) indicating the internal placement of high dose rate treatment catheters.

investigated the use of MEX printing based on CT images [163, 164] or optical surface scanning and photogrammetry [146, 165] to produce geometrically accurate and physically robust applicators [165] that minimise air gaps and improve stability and source positioning reproducibility [163, 166], while also maintaining approximate water equivalence [163, 166].

There is an unprecedented level of control achievable by designing virtual applicators for 3D printing rather than manually moulding applicators from wax or thermoplastics using plaster casts or directly onto the patient's skin [162, 165]. The ability to exactly control applicator thicknesses and catheter positions means that the planned dose can be optimised ahead of printing, to achieve an optimal dose distribution [143, 164, 165], even when the targeted lesion varies in depth [167]. Illustrations showing designs for 3D-printed superficial brachytherapy applicators are shown in figure 9.12.

9.6.3 Interstitial templates

Interstitial brachytherapy is a well-established technique where needles or catheters are used to place the brachytherapy source within soft tissue targets such as prostate, breast, or tongue. Interstitial brachytherapy has also been used to deliver high dose rate (HDR) treatments to challenging targets such as liver metastases and brain cancers [168, 169]. Generally, in interstitial treatments, a 3D dose distribution is shaped to cover the tumour volume using either HDR or low dose rate (LDR) brachytherapy (see chapter 1).

Interstitial HDR brachytherapy treatments involve the programmed placement of a high activity source (e.g. iridium-192) at a series of dwell positions within an array of catheters (usually needles). Treatments are fractionated over a small number of days, between which the needles may be kept *in situ* or removed from the patient and re-inserted. The array of needles must be kept stable throughout the treatment, with all needle positions kept constant with respect to all other needle positions and with respect to the patient's anatomy.

As described in chapter 1, LDR brachytherapy treatments are delivered using permanently implanted radioactive seeds. LDR seeds are placed using needles that are removed after seed insertion, but which must nonetheless be kept stable throughout seed insertion.

For both HDR and LDR brachytherapy, the key task of stabilising the needle placement is achieved using a template located externally to the patient, holding the needles in place close to the surface where they enter the body. Conventionally, interstitial brachytherapy templates are designed to keep needles straight, parallel and evenly spaced. However, the advent of 3D printing has led to investigations of alternative needle arrangements using patient-specific templates, where needle positions are planned in advance of implantation to achieve optimal target coverage [158, 159] and needle arrangements can be deliberately non-parallel (called non-coplanar) in order to treat complex targets while sparing critical structures or avoiding bones in the needle path [142]. Figure 9.13 gives a clinical example of a 3D-printed noncoplanar HDR brachytherapy needle template [142].

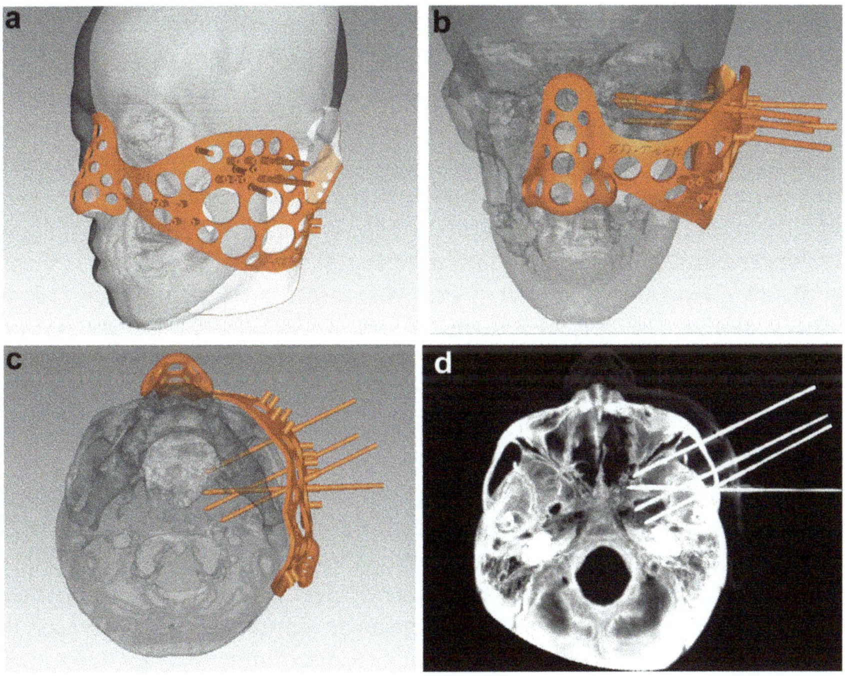

Figure 9.13. 'The appearance of digital stereotactic model on three-quarter view (a), anterior view (b), and inferior view (c). The actual position of needles and individual template were in accordance with digital stereotactic model (d)' [142]. Reprinted from [142], Copyright (2019), with permission from Elsevier.

Imaging the patient up to two days before surgery [158] allows the interstitial brachytherapy treatment to be pre-planned, with needle positions carefully arranged and adjusted to meet treatment goals [142]. The interstitial template can then be designed to achieve the planned needle positions, with either a thick template with angled holes indicating needle positions [150] or a thin layer fitted to the patient's external contour, and needle positions and directions indicated using small hollow cylinders in the surface through which the needles may be implanted [142, 157]. Repeated imaging during the implantation process to verify correct alignment of the template, correct position and depth of all needles, and for LDR correct positioning of implanted seeds, is used to confirm that the brachytherapy treatments can and will be delivered accurately [142, 151, 158]. Nonetheless, separate HDR brachytherapy treatment planning scans and LDR post-implant dosimetry scans remain important for patient safety.

9.6.4 Intracavitary moulds

Brachytherapy treatment of tumours within bodily cavities such as the vagina and anal canal are traditionally done using commercially available generic applicators such as cylinders. While these cylinders are available in various sizes, cavities of the body are rarely cylindrical. Variations due to surgery, external beam radiation therapy, or tumour size may introduce air gaps between the applicator and the cavity, reducing dose coverage [170]. In extreme examples, the cavity may no longer be able to accommodate the applicator at all.

An established method of solving these challenging cases is to produce a patient-specific mould using, for example, a silicone-based material [141, 171]. These silicone moulds are able to minimise air gaps between the applicator and cavity surface, while offering the freedom of catheter placement within the applicator. The resulting dosimetry is potentially more conformal to the tumour geometry, but reliability is limited by the material's non-water equivalence [172] and the imprecise placement of the catheters during the mould's manufacturing process [173].

Customised 3D-printed applicators provide a more water-equivalent [17, 44, 166] and precise alternative for treatment of intracavitary brachytherapy patients, regardless of any geometrical challenges (see figure 9.14). 3D simulation of the cavity to be treated (e.g. CT or MRI imaging) allows for the virtual design of patient-specific applicators [5, 147], where channel positions can be informed by optimised dose distributions. The image guided design also provides the accuracy to plunge the needles past the applicator's surface and into the cavity wall, providing an interstitial component to treat deeper tumours [19, 174].

The level of regulatory approval and sterilisation required by 3D-printed medical devices is dependent on the risks of using the device. For example, an intracavitary mould may require a lesser level of sterilisation compared to a device used for surgical needle insertion. Consultation with local regulatory bodies is recommended when manufacturing any medical device (also see section 9.2 and chapter 7).

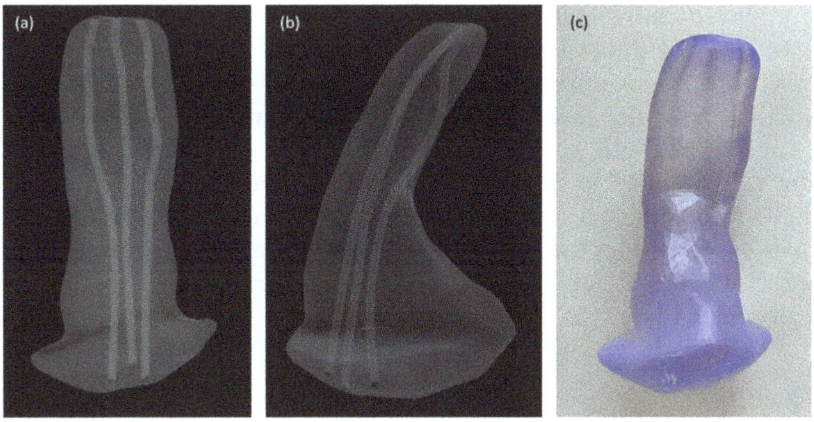

Figure 9.14. (a) and (b) images of STL models and (c) photograph of a 3D-printed photopolymer resin applicator for vaginal intracavitary brachytherapy, showing external shape created to match patient anatomy and catheter channel positions optimised to achieve prescribed dose throughout irregularly shaped tumour.

9.6.5 Dose calculation considerations

Historically, the main brachytherapy dose calculation algorithm used in radiation therapy centres has been the AAPM TG 43 formalism [175], which assumed that the treatment equipment and patient are water-equivalent. For users of this formalism, it is therefore necessary to assess the water-equivalence of all applicators in clinical use and thereby evaluate the accuracy with which dose distributions can be represented by the treatment planning software.

Water equivalence has been investigated for common materials used in 3D printing including MEX thermoplastics [17, 44, 166], SLA photopolymers [5], and SLS polyamides [176]. Generally, 3D-printing plastics appear to be mostly water equivalent at brachytherapy energies, though it remains important to verify this for each new material considered for clinical use. For example, a silicone-based material can lead to an underestimate of dose from 4% [172] to 10% [141] if material characteristics are ignored under the AAPM TG 43 formalism.

Superficial treatment sites introduce another source of uncertainty in dose calculations using the TG 43 formalism [143, 162]. The assumption that the entire calculation volume is water applies both inside and outside the patient, leading to an overestimate of backscattered dose delivered to the target. 3D-printing custom applicators provide the opportunity to rectify the absence of this scatter material, thereby increasing the accuracy of the clinical dose calculation algorithm.

Clearly, if there is a clinical motivation for using non-water equivalent materials in brachytherapy treatments, then an appropriately commissioned and verified model-based dose calculation algorithm (MBDCA) should be used for treatment planning [177]. Applicator manufacturers often supply virtual models for use with MBDCAs, providing both the geometric and material properties of their applicator in the model, although it may be more challenging for individual users to specify or select appropriate settings for new materials—potentially leading to another source of uncertainty in the dose calculation.

Nonetheless, a major appeal of MBDCA adoption for brachytherapy treatment planning is the potential to design and use more complex patient-specific brachytherapy applicators, with greater accuracy and safety than was achievable when everything was assumed to be water. The virtual modelling of 3D-printed devices based on 3D medical images, within a sophisticated treatment planning system that uses an MBDCA, enables further customisation of the dose distribution by utilising integrated dose modulation via infill (i.e. density) variations [44, 178] or multi-material printing [15].

9.7 Key points

- Patient safety concerns related to 3D-printed radiation therapy devices can be addressed by adhering to stringent national safety standards, even from outside the jurisdiction covered by those standards, and by creating and following a well-documented quality management system (see chapter 7).
- 3D-printed boluses, compensators, and range shifters have advantages over non-3D-printed alternatives in terms of patient fit, comfort, published treatment outcomes (for photons), patient time spent in the radiation therapy centre, and time needed for treatment setup.
- 3D-printed radiation shields can replace the use of lead in TSET, and have the potential to replace lead in other electron or kV photon treatments.
- 3D printing can also be used to improve the reliability of fabricating LMPA cutouts for electron radiation therapy treatments, or to eliminate the use of LMPA altogether.
- 3D-printed immobilisation devices have been used to improve patient comfort during head-and-neck radiation therapy treatments, and to improve immobilisation and organ-at-risk sparing for challenging radiation therapy treatments of the extremities and oral cavity.
- Patient-specific 3D-printed superficial applicators, interstitial needle templates, and intracavity moulds can improve the quality of brachytherapy treatments for challenging anatomical sites and extend the range of pathologies that can be treated using this modality.

References

[1] Kairn T 2020 Applying good software development processes in practice *Diagnostic Radiology Physics with MATLAB* (Boca Raton, FL: CRC Press) pp 129–40

[2] Kudchadker R, Antolak J, Morrison W, Wong P and Hogstrom K 2003 Utilization of custom electron bolus in head and neck radiotherapy *J. Appl. Clin. Med. Phys.* **4** 321–33

[3] Park J and Yea J 2016 Three-dimensional customized bolus for intensity-modulated radiotherapy in a patient with kimura's disease involving the auricle *Cancer/Radiothérapie* **20** 205–9

[4] Cleland S *et al* 2021 Dosimetric evaluation of a patient-specific 3D-printed oral positioning stent for head-and-neck radiotherapy *Phys. Eng. Sci. Med.* **44** 887–99

[5] Wiebe E, Easton H, Thomas G, Barbera L, D'Alimonte L and Ravi A 2015 Customized vaginal vault brachytherapy with computed tomography imaging-derived applicator prototyping *Brachytherapy* **14** 380–4

[6] Lancellotta V *et al* 2019 Individual 3-dimensional printed mold for treating hard palate carcinoma with brachytherapy: a clinical report *J. Prosthet. Dentistry* **121** 690–3

[7] Morrison W H, Wong P-F, Starkschall G, Garden A S, Childress C, Hogstrom K R and Peters L J 1995 Water bolus for electron irradiation of the ear canal *Int. J. Radiat. Oncol. Biol. Phys.* **33** 479–83

[8] Standards Australia 1999 *Plastics Materials for Food Contact Use (AS 2070-1999)* (Sydney: Standards Australia)

[9] Rankin T M, Giovinco N A, Cucher D J, Watts G, Hurwitz B and Armstrong D G 2014 Three-dimensional printing surgical instruments: are we there yet? *J. Surgical Res.* **189** 193–7

[10] International Organization for Standardization 2018 *Iso 10 993-1—Biological Evaluation of Medical Devices—Part 1: Evaluation and Testing Within a Risk Management Process* [Online] Available: https://www.iso.org/standard/68936.html (visited on 06/01/2023)

[11] Alifui-Segbaya F, Bowman J, White A R, George R, Fidan I and Love R M 2020 Chemical characterization of additively manufactured methacrylates for dental devices *Additive Manuf* **31** 100944

[12] Rogers H B, Zhou L T, Kusuhara A, Zaniker E, Shafaie S, Owen B C, Duncan F E and Woodruff T K 2021 Dental resins used in 3D printing technologies release ovo-toxic leachates *Chemosphere* **270** 129003

[13] Standards Australia/Standards New Zealand 2014 *Australian/New Zealand Standard TM Reprocessing of Reusable Medical Devices in Health Service Organizations (AS/NZS 4187:2014)* (Sydney: Standards Australia/Standards New Zealand)

[14] Perez M, Block M, Espalin D, Winker R, Hoppe T, Medina F and Wicker R 2012 Sterilization of FDM-manufactured parts *2012 Int. Solid Freeform Fabrication Symp.* (Austin, TX: University of Texas at Austin)

[15] Skinner L B, Niedermayr T, Prionas N, Perl J, Fahimian B and Kidd E A 2020 Intensity modulated Ir-192 brachytherapy using high-Z 3D printed applicators *Phys. Med. Biol.* **65** 155018

[16] D'Alimonte L, Ravi A, Helou J, Morrison D, Mendez L C, Easton H and Morton G 2019 Optimized penile surface mold brachytherapy using latest stereolithography techniques: a single-institution experience *Brachytherapy* **18** 348–52

[17] Cunha J, Mellis K, Sethi R, Siauw T, Sudhyadhom A, Garg A, Goldberg K, Hsu I and Pouliot J 2015 Evaluation of PC-ISO for customized, 3D printed, gynecologic 192ir HDR *J. Appl. Clin. Med. Phys.* **16** 246–53

[18] Sethi R, Cunha A, Mellis K, Siauw T, Diederich C, Pouliot J and Hsu I-C 2016 Clinical applications of custom-made vaginal cylinders constructed using three-dimensional printing technology *J. Contemp. Brachyther* **3** 208–14

[19] Lindegaard J C, Madsen M L, Traberg A, Meisner B, Nielsen S K, Tanderup K, Spejlborg H, Fokdal L U and Nørrevang O 2016 Individualised 3D printed vaginal template for MRI guided brachytherapy in locally advanced cervical cancer *Radiother. Oncol.* **118** 173–5

[20] Wilks R, Skewes J, Forrestal D, Kairn T and Crowe S 2023 Evaluating 3D printing materials for intracavitary gynaecological brachytherapy *Phys. Eng. Sci. Med.* **46** 482

[21] Mohan R and Grosshans D 2017 Proton therapy–present and future *Adv. Drug. Deliv. Rev.* **109** 26–44

[22] Mohan R, Das I J and Ling C C 2017 Empowering intensity modulated proton therapy through physics and technology: an overview *Int. J. Radiat. Oncol. Biol. Phys.* **99** 304–16

[23] Ju S G, Kim M K, Hong C-S, Kim J S, Han Y, Choi D H, Shin D and Lee S B 2014 New technique for developing a proton range compensator with use of a 3-dimensional printer *Int. J. Radiat. Oncol. Biol. Phys.* **88** 453–8

[24] Zou W *et al* 2015 Potential of 3D printing technologies for fabrication of electron bolus and proton compensators *J. Appl. Clin. Med. Phys.* **16** 90–8

[25] Michiels S, Barragán A M, Souris K, Poels K, Crijns W, Lee J A, Sterpin E, Nuyts S, Haustermans K and Depuydt T 2018 Patient-specific bolus for range shifter air gap reduction in intensity-modulated proton therapy of head-and-neck cancer studied with Monte Carlo based plan optimization *Radiother. Oncol.* **128** 161–6

[26] Wochnik A, Stolarczyk L and Ambrožová I *et al* 2021 Out-of-field doses for scanning proton radiotherapy of shallowly located paediatric tumours—a comparison of range shifter and 3D printed compensator *Phys. Med. Biol.* **66** 035012

[27] Rooney M K, Rosenberg D M, Braunstein S, Cunha A, Damato A L, Ehler E, Pawlicki T, Robar J, Tatebe K and Golden D W 2020 Three-dimensional printing in radiation oncology: a systematic review of the literature *J. Appl. Clin. Med. Phys.* **21** 15–26

[28] Albantow C E and Brown S J 2022 Three-dimensional printer use in the Australian and New Zealand radiation therapy setting *J. Med. Radiat. Sci.* **69** 492–501

[29] Tino R, Leary M, Yeo A, Kyriakou E, Kron T and Brandt M 2020 Additive manufacturing in radiation oncology: a review of clinical practice, emerging trends and research opportunities *Int. J. Extreme Manuf.* **2** 012003

[30] Vyas V, Palmer L, Mudge R, Jiang R, Fleck A, Schaly B, Osei E and Charland P 2013 On bolus for megavoltage photon and electron radiation therapy *Med. Dosim.* **38** 268–73

[31] Benoit J, Pruitt A F and Thrall D E 2009 Effect of wetness level on the suitability of wet gauze as a substitute for superflab as a bolus material for use with 6 MV photons *Veterinary Radiol. Ultrasound* **50** 555–9

[32] Zhao Y, Moran K, Yewondwossen M, Allan J, Clarke S, Rajaraman M, Wilke D, Joseph P and Robar J L 2017 Clinical applications of 3-dimensional printing in radiation therapy *Med. Dosim.* **42** 150–5

[33] Burleson S, Baker J, Hsia A T and Xu Z 2015 Use of 3D printers to create a patient-specific 3D bolus for external beam therapy *J. Appl. Clin. Med. Phys.* **16** 166–78

[34] Ehler E, Sterling D, Dusenbery K and Lawrence J 2018 Workload implications for clinic workflow with implementation of three-dimensional printed customized bolus for radiation therapy: a pilot study *PLOS One* **13** e0204944

[35] Park S-Y, Choi C H, Park J M, Chun M, Han J H and Kim J i 2016 A patient-specific polylactic acid bolus made by a 3D printer for breast cancer radiation therapy *PLOS One* **11** e0168063

[36] Canters R A, Lips I M, Wendling M, Kusters M, van Zeeland M, Gerritsen R M, Poortmans P and Verhoef C G 2016 Clinical implementation of 3D printing in the construction of patient specific bolus for electron beam radiotherapy for non-melanoma skin cancer *Radiother. Oncol.* **121** 148–53

[37] Robar J L, Moran K, Allan J, Clancey J, Joseph T, Chytyk-Praznik K, MacDonald R L, Lincoln J, Sadeghi P and Rutledge R 2018 Intrapatient study comparing 3D printed bolus

versus standard vinyl gel sheet bolus for postmastectomy chest wall radiation therapy *Pract. Radiat. Oncol.* **8** 221–9
[38] Hsu E J, Parsons D, Chiu T, Godley A R, Sher D J and Vo D T 2022 3D printed integrated bolus/headrest for radiation therapy for malignancies involving the posterior scalp and neck 3D Print *3D Print. Med.* **8** 1–7
[39] Kairn T *et al* 2021 Determining tolerance levels for quality assurance of 3D printed bolus for modulated arc radiotherapy of the nose *Phys. Eng. Sci. Med.* **44** 1187–99
[40] Asfia A, Novak J I, Mohammed M I, Rolfe B and Kron T 2020 A review of 3D printed patient specific immobilisation devices in radiotherapy *Phys. Imaging Radiat. Oncol.* **13** 30–5
[41] Lukowiak M, Jezierska K, Boehlke M, Wiecko M, £ukowiak A, Podraza W, Lewocki M, Masojae B and Falco M 2017 Utilization of a 3D printer to fabricate boluses used for electron therapy of skin lesions of the eye canthi *J. Appl. Clin. Med. Phys.* **18** 76–81
[42] Van der Walt M, Crabtree T and Albantow C 2019 PLA as a suitable 3D printing thermoplastic for use in external beam radiotherapy *Australas. Phys. Eng. Sci. Med.* **42** 1165–76
[43] Kairn T, Crowe S and Markwell T 2015 Use of 3D printed materials as tissue-equivalent phantoms *World Congress on Medical Physics and Biomedical Engineering, June 7–12, 2015, Toronto, Canada* (Berlin: Springer) pp 728–31
[44] Ricotti R *et al* 2016 3d-printed applicators for high dose rate brachytherapy: dosimetric assessment at different infill percentage *Phys. Med.* **32** 1698–706
[45] Kairn T, Zahrani M, Cassim N, Livingstone A, Charles P and Crowe S 2020 Quasi-simultaneous 3D printing of muscle-, lung-and bone-equivalent media: a proof-of-concept study *Phys. Eng. Sci. Med.* **43** 701–10
[46] Kairn T, Poroa T, Cleland S, Dawes J, Kenny L, Lancaster C, Lin C and Crowe S 2023 Clinical use of 3D printed intra-oral stents: opportunities and obstacles *Phys. Eng. Sci. Med.* **46** 480–1
[47] Wills R J, Lowe G, Jones C L and Hoskin P J 2021 Dosimetric comparison of volumetric modulated arc therapy (VMAT) and high-dose-rate brachytherapy (HDR-BT) for superficial skin irradiation with significant curvature in one or more planes *Strahlentherapie und Onkologie* **197** 547–54
[48] Chua B, Jackson J E, Lin C and Veness M J 2019 Radiotherapy for early non-melanoma skin cancer *Oral Oncol.* **98** 96–101
[49] Spelleken E, Kairn T, Crowe S and Hellyer J 2022 Effects of air gaps when using 3D printed bolus for small field VMAT treatments *Phys. Eng. Sci. Med.* **45** 351–2
[50] Kairn T, Poroa T, Luscombe J, Chan S, Grogan M and Crowe S B 2023 Patient-specific immobilisation for radiotherapy treatment of extensive lower-limb carcinoma *IFMBE Proc.* (in press)
[51] Anderson P R, Hanlon A L, McNeeley S W and Freedman G M 2004 Low complication rates are achievable after postmastectomy breast reconstruction and radiation therapy *Int. J. Radiat. Oncol. Biol. Phys.* **59** 1080–7
[52] Zhang Y, Huang Y, Ding S, Liang J, Kuang J, Mao Q, Ying W, Shu Y, Li J and Jiang C 2022 A clinical trial to compare a 3D-printed bolus with a conventional bolus with the aim of reducing cardiopulmonary exposure in postmastectomy patients with volumetric modulated arc therapy *Cancer Med.* **11** 1037–47
[53] Sasaki D K, McGeachy P, Aviles J E A, McCurdy B, Koul R and Dubey A 2019 A modern mold room: Meshing 3D surface scanning, digital design, and 3D printing with bolus fabrication *J. Appl. Clin. Med. Phys.* **20** 78–85

[54] Yu S, Ahn S H, Choi S H, Ahn W S and Jung I-h 2023 Clinical application of a customized 3D-printed bolus in radiation therapy for distal extremities *Life* **13** 362

[55] Chiu T, Tan J, Brenner M, Gu X, Yang M, Westover K, Strom T, Sher D, Jiang S and Zhao B 2018 Three-dimensional printer-aided casting of soft, custom silicone boluses (SCSBs) for head and neck radiation therapy *Pract. Radiat. Oncol.* **8** e167–74

[56] Kim S-W, Shin H-J, Kay C S and Son S H 2014 A customized bolus produced using a 3-dimensional printer for radiotherapy *PLOS One* **9** e110746

[57] Kim S-W, Kwak J, Cho B, Song S Y, Lee S-W and Jeong C 2017 Clinical implementation of 3D printing in the construction of patient specific bolus for photon beam radiotherapy for mycosis fungoides Prog *Med. Phys.* **28** 33–8

[58] Low D, Starkschall G, Bujnowski S, Wang L and Hogstrom K 1992 Electron bolus design for radiotherapy treatment planning: Bolus design algorithms *Med. Phys.* **19** 115–24

[59] Perkins G H, McNeese M D, Antolak J A, Buchholz T A, Strom E A and Hogstrom K R 2001 A custom three-dimensional electron bolus technique for optimization of postmastectomy irradiation *Int. J. Radiat. Oncol. Biol. Phys.* **51** 1142–51

[60] Su S, Moran K and Robar J L 2014 Design and production of 3D printed bolus for electron radiation therapy *J. Appl. Clin. Med. Phys.* **15** 194–211

[61] Lee V W, Liu A C, Cheng K W, Chiang C-L and Lee V H-F 2023 Dosimetric benefits of 3D-printed modulated electron bolus following lumpectomy and whole-breast radiotherapy for left breast cancer *Med. Dosim.* **48** 37–43

[62] Lukowiak M, Boehlke M, Matias D, Jezierska K, Piatek-Hnat M, Lewocki M, Podraza W, El Fray M and Kot W 2016 Use of a 3D printer to create a bolus for patients undergoing tele-radiotherapy Int *J. Radiat. Res.* **14** 287–95

[63] Zeidan O A, Chauhan B D, Estabrook W W, Willoughby T R, Manon R R and Meeks S L 2011 Image-guided bolus electron conformal therapy–a case study *J. Appl. Clin. Med. Phys.* **12** 68–75

[64] Khan F M and Gibbons J P 2014 *Khan's the Physics of Radiation Therapy* (Baltimore, MD: Williams & Wilkins)

[65] Kong M and Holloway L 2007 An investigation of central axis depth dose distribution perturbation due to an air gap between patient and bolus for electron beams *Australas. Phys. Eng. Sci. Med.* **30** 111–9

[66] Lindsay C, Kumlin J, Jirasek A, Lee R, Martinez D M, Schaffer P and Hoehr C 2015 3D printed plastics for beam modulation in proton therapy *Phys. Med. Biol.* **60** N231–40

[67] Lindsay C, Kumlin J, Martinez D, Jirasek A and Hoehr C 2016 Design and application of 3D-printed stepless beam modulators in proton therapy *Phys. Med. Biol.* **61** N276

[68] Huang S, Kang M, Souris K, Ainsley C, Solberg T D, McDonough J E, Simone C B and Lin L 2018 Validation and clinical implementation of an accurate Monte Carlo code for pencil beam scanning proton therapy *J. Appl. Clin. Med. Phys.* **19** 558–72

[69] Shen J, Liu W, Anand A, Stoker J B, Ding X, Fatyga M, Herman M G and Bues M 2015 Impact of range shifter material on proton pencil beam spot characteristics *Med. Phys.* **42** 1335–40

[70] Bertschi S, Krieger M, Weber D C, Lomax A J and van de Water S 2022 Impact of spot reduction on the effectiveness of rescanning in pencil beam scanned proton therapy for mobile tumours *Phys. Med. Biol.* **67** 215019

[71] Michiels S *et al* 2016 Towards 3D printed multifunctional immobilization for proton therapy: initial materials characterization *Med. Phys.* **43** 5392–402

[72] Shirey R J and Wu H T 2018 Quantifying the effect of air gap, depth, and range shifter thickness on TPS dosimetric accuracy in superficial pbs proton therapy *J. Appl. Clin. Med. Phys.* **19** 164–73

[73] Crowe S B, Bennett J, Lathouras M, Lancaster C M, Sylvander S R, Chua B, Bettington C S, Lin C Y and Kairn T 2020 Impact of radiopacified bone cement on radiotherapy dose calculation *Phys. Imaging Radiat. Oncol.* **14** 12–6
[74] Kairn T, Buddhavarapu A, Cassim N, Charles P, Jessen L and Crowe S 2022 Lessons learnt on the road to fabricating materials that are bone-equivalent in both kilovoltage and megavoltage photon beams *Phys. Eng. Sci. Med.* **45** 308
[75] Vozenin M-C, Bourhis J and Durante M 2022 Towards clinical translation of FLASH radiotherapy *Nat. Rev. Clin. Oncol.* **19** 1–13
[76] Mascia A E *et al* 2023 Proton FLASH radiotherapy for the treatment of symptomatic bone metastases: the FAST-01 nonrandomized trial *JAMA Oncol.* **9** 62–9
[77] Simeonov Y, Weber U, Penchev P, Ringbæk T P, Schuy C, Brons S, Engenhart-Cabillic R, Bliedtner J and Zink K 2017 3d range-modulator for scanned particle therapy: development, Monte Carlo simulations and experimental evaluation *Phys. Med. Biol.* **62** 7075
[78] Simeonov Y, Weber U, Schuy C, Engenhart-Cabillic R, Penchev P, Flatten V and Zink K 2022 Development, Monte Carlo simulations and experimental evaluation of a 3D range-modulator for a complex target in scanned proton therapy *Biomed. Phys. Eng. Express* **8** 035006
[79] Wani A L, Ara A and Usmani J A 2015 Lead toxicity: a review *Interdiscip. Toxicol.* **8** 55–64
[80] Kosnett M J, Wedeen R P, Rothenberg S J, Hipkins K L, Materna B L, Schwartz B S, Hu H and Woolf A 2007 Recommendations for medical management of adult lead exposure *Environ. Health Perspect.* **115** 463–71
[81] Wang M, Chen Z, Song W, Hong D, Huang L and Li Y 2021 A review on cadmium exposure in the population and intervention strategies against cadmium toxicity *Bull. Environ. Contamination Toxicol.* **106** 65–74
[82] Soylu H, Yurt Lambrecht F and Ersöz O 2015 Gamma radiation shielding efficiency of a new lead-free composite material *J. Radioanal. Nucl. Chem.* **305** 529–34
[83] Yaffe M J, Mawdsley G E, Lilley M, Servant R and Reh G 1991 Composite materials for x-ray protection *Health Phys.* **60** 661–4
[84] Kairn T, Maxwell S, Trapp J and Crowe S 2021 Assessment of integrity and lead-equivalence of shielded garments using two-dimensional x-ray images from a computed tomography scanner *Radiat. Prot. Dosim.* **193** 155–64
[85] Eid A and Zawia N 2016 Consequences of lead exposure, and its emerging role as an epigenetic modifier in the aging brain *NeuroToxicology* **56** 254–61
[86] Yan H, Guo F, Zhu D, Stryker S, Trumpore S, Roberts K, Higgins S, Nath R, Chen Z and Liu W 2017 On the use of bolus for pacemaker dose measurement and reduction in radiation therapy *J. Appl. Clin. Med. Phys.* **19** 125–31
[87] Jia S B, Soleimani A, Mirsadraee M, Zarifi S and Sanaeifar E 2023 Evaluation of the effectiveness of testicular shielding in rectal cancer radiotherapy *Radiat. Phys. Chem.* **202** 110435
[88] Luis S, Christie D, Kaminski A, Kenny L and Peres M 2009 Pregnancy and radiotherapy: management options for minimising risk, case series and comprehensive literature review *J. Med. Imaging Radiat. Oncol.* **53** 559–68
[89] Leat C J, Ramachandran P and Langton C M 2020 Tungsten carbide and LMPA electron cutouts: comparison and validation using Monte Carlo modelling and measurement of dose *Biomed. Phys. Eng. Express* **7** 015007
[90] Elmahmoud Z, Gunther J R and Christopherson K 2023 A simple solution to create a custom scalp-sparing helmet to prevent alopecia in patients undergoing total skin electron beam therapy for cutaneous T cell lymphoma *Clin. Transl. Radiat. Oncol.* **38** 53–6

[91] Kairn T, Wilks R, Yu L, Lancaster C and Crowe S B 2020 In vivo monitoring of total skin electron dose using optically stimulated luminescence dosimeters *Rep. Pract. Oncol. Radiother.* **25** 35–40

[92] Rahimy E, Skinner L, Kim Y H and Hoppe R T 2021 Technical report: 3D-printed patient-specific scalp shield for hair preservation in total skin electron beam therapy *Tech. Innov. Patient Support Radiat. Oncol.* **18** 12–5

[93] Choi C, Kim J and Park J 2019 Feasibility of metal testicle shield by 3D printing for CT image-based treatment planning *Int. J. Radiat. Oncol. Biol. Phys.* **105** E684

[94] Burke S E, Kassaee A, Plastaras J P, Gray A, Hill-Kayser C E, Ding X, Thorne N, Teo B-K K and LaRiviere M J 2022 Applications of a CT-compatible 3D printed plastic scrotal 'clamshell' for radiotherapy planning, proton beam shielding, and testicular immobilization *J. Case Rep. Med.* **11** 48–54

[95] Sharma A *et al* 2018 Low-cost optical scanner and 3-dimensional printing technology to create lead shielding for radiation therapy of facial skin cancer: first clinical case series *Adv. Radiat. Oncol.* **3** 288–96

[96] Crowe S B, Charles P H, Cassim N, Maxwell S K, Sylvander S R, Smith J G and Kairn T 2021 Predicting the required thickness of custom shielding materials in kilovoltage radiotherapy beams *Phys. Med.* **81** 94–101

[97] Michiels S, Mangelschots B, Roover R D, Devroye C and Depuydt T 2018 Production of patient-specific electron beam aperture cut-outs using a low-cost, multi-purpose 3D printer *J. Appl. Clin. Med. Phys.* **19** 756–60

[98] Moore M J, Snelgrove R, Darko J and Osei E K 2022 One layer at a time: the use of 3D printing in the fabrication of cadmium-free electron field shaping devices *J. Radiother. Pract.* **21** 222–7

[99] Skinner L, Fahimian B P and Yu A S 2019 Tungsten filled 3D printed field shaping devices for electron beam radiation therapy *PLOS One* **14** e0217757

[100] Breitkreutz D Y, Skinner L, Lo S and Yu A 2021 Nontoxic electron collimators *J. Appl. Clin. Med. Phys.* **22** 73–81

[101] Cheng J C, Dubey A, Beck J, Sasaki D, Leylek A and Rathod S 2022 Optical scan and 3D printing guided radiation therapy—an application and provincial experience in cutaneous nasal carcinoma *3D Print. Med.* **8** 8

[102] Briggs M, Clements H, Wynne N, Rennie A and Kellett D 2016 3D printed facial laser scans for the production of localised radiotherapy treatment masks–a case study *J. Vis. Commun. Med.* **39** 99–104

[103] Vassiliou M, Callender J and Manning-Stanley A 2019 An evaluation of techniques used in superficial radiotherapy for non-melanoma skin cancer to replicate the planned treatment area: a prospective study *Radiography* **25** 280–7

[104] Crowe S *et al* 2022 Evaluation of optical 3D scanning system for radiotherapy use *J. Med. Radiat. Sci.* **69** 218–26

[105] Verhey L J 1995 Immobilizing and positioning patients for radiotherapy *Semin. Radiat. Oncol.* **5** 100–14

[106] Tanikella N G, Wittbrodt B and Pearce J M 2017 Tensile strength of commercial polymer materials for fused filament fabrication 3D printing *Additive Manuf.* **15** 40–7

[107] Mostafa K G, Montemagno C and Qureshi A J 2018 Strength to cost ratio analysis of FDM Nylon 12 3D printed parts *Proc. Manuf.* **26** 753–62

[108] Asfia A, Deepak B, Novak J I, Rolfe B and Kron T 2020 Infill selection for 3D printed radiotherapy immobilisation devices *Biomed. Phys. Eng. Express* **6** 065014

[109] Dialami N, Chiumenti M, Cervera M, Rossi R, Chasco U and Domingo M 2021 Numerical and experimental analysis of the structural performance of am components built by fused filament fabrication *Int. J. Mech. Mater. Des.* **17** 225–44

[110] Wittbrodt B and Pearce J M 2015 The effects of PLA color on material properties of 3-D printed components *Additive Manuf.* **8** 110–6

[111] Lu L, Sharf A, Zhao H, Wei Y, Fan Q, Chen X, Savoye Y, Tu C, Cohen-Or D and Chen B 2014 Build-to-last: Strength to weight 3D printed objects *ACM Trans. Graph.* **33** 1–10

[112] Stava O, Vanek J, Benes B, Carr N and Mìch R 2012 Stress relief: improving structural strength of 3D printable objects *ACM Trans. Graph.* **31** 1–11

[113] Baltz G C, Briere T, Luo D, Howell R M, Krafft S and Han E Y 2020 3D-printed headrest for frameless Gamma Knife radiosurgery: Design and validation *J. Appl. Clin. Med. Phys.* **21** 6–15

[114] Kelly A, Hardcastle N, Metcalfe P, Cutajar D, Quinn A, Foo K, Cardoso M, Barlin S and Rosenfeld A 2011 Surface dosimetry for breast radiotherapy in the presence of immobilization cast material *Phys. Med. Biol.* **56** 1001

[115] Navarro-Martin A, Cacicedo J, Leaman O, Sancho I, García E, Navarro V and Guedea F 2015 Comparative analysis of thermoplastic masks versus vacuum cushions in stereotactic body radiotherapy *Radiat. Oncol.* **10** 1–8

[116] Bolsi A, Lomax A J, Pedroni E, Goitein G and Hug E 2008 Experiences at the Paul Scherrer Institute with a remote patient positioning procedure for high-throughput proton radiation therapy *Int. J. Radiat. Oncol. Biol. Phys.* **71** 1581–90

[117] Zheng X, Dai T, Shu X, Pu Y, Feng G, Li X, Liao D and Du X 2012 A new method of lower extremity immobilization in radiotherapy *Radiat. Oncol.* **7** 27

[118] Leech M, Coffey M, Mast M, Moura F, Osztavics A, Pasini D and Vaandering A 2017 ESTRO ACROP guidelines for positioning, immobilisation and position verification of head and neck patients for radiation therapists *Tech. Innov. Patient Support Radiat. Oncol.* **1** 1–7

[119] Miron V M, Etzelstorfer T, Kleiser R, Raffelsberger T, Major Z and Geinitz H 2022 Evaluation of novel 3D-printed and conventional thermoplastic stereotactic high-precision patient fixation masks for radiotherapy *Strahlentherapie und Onkologie* **198** 1032–41

[120] Haefner M F *et al* 2018 3D-printed masks as a new approach for immobilization in radiotherapy—a study of positioning accuracy *Oncotarget* **9** 6490

[121] Mattke M, Rath D, Häfner M, Unterhinninghofen R, Sterzing F, Debus J and Giesel F L 2021 Individual 3d-printed fixation masks for radiotherapy: first clinical experiences *Int. J. Comput. Assist. Radiol. Surg.* **16** 1043–9

[122] Nixon J L, Brown B, Pigott A E, Turner J, Brown E, Bernard A, Wall L R, Ward E C and Porceddu S V 2019 A prospective examination of mask anxiety during radiotherapy for head and neck cancer and patient perceptions of management strategies *J. Med. Radiat. Sci.* **66** 184–90

[123] Fisher M, Applegate C, Ryalat M, Laycock S, Hulse M, Emmens D and Bell D 2014 Evaluation of 3-D printed immobilisation shells for head and neck IMRT *Open J. Radiol.* **4** 322–8

[124] Loja M, Craveiro D, Vieira L, Sousa E, Rodrigues J and Portal R 2019 Radiotherapy-customized head immobilization masks: from modeling and analysis to 3D printing *Nucl. Sci. Tech.* **30** 1–16

[125] Robar J L *et al* 2022 Novel multi jet fusion 3D-printed patient immobilization for radiation therapy *J. Appl. Clin. Med. Phys.* **23** e13773

[126] Asfia A, Novak J I, Rolfe B and Kron T 2022 Development of a patient-specific immobilisation facemask for radiation therapy using additive manufacturing, pressure sensors and topology optimisation *Rapid Prototyping J.* **28** 945–52

[127] Pham Q-V, Lavallée A P, Foias A, Roberge D, Mitrou E and Wong P 2018 Radiotherapy immobilization mask molding through the use of 3D-printed head models *Technol. Cancer Res. Treat.* **17** 1533033818809051

[128] Chen C-P *et al* 2022 Skin surface dose for whole breast radiotherapy using personalized breast holder: comparison with various radiotherapy techniques and clinical experiences *Cancers* **14** 3205

[129] Arimura T, Ogino T, Yoshiura T, Matsuyama M, Kondo N, Miyazaki H, Sakuragi A, Ohara T, Ogo E and Hishikawa Y 2017 A feasibility study of a hybrid breast-immobilization system for early breast cancer in proton beam therapy *Med. Phys.* **44** 1268–74

[130] Denton A S and Maher J 2003 Interventions for the physical aspects of sexual dysfunction in women following pelvic radiotherapy *Cochrane Database Syst. Rev.* **2015** CD003750

[131] Perez B A, Willett C G, Czito B G and Palta M 2015 Anal canal cancer *Intensity-Modulated Radiation Therapy Clinical Evidence and Techniques* (Tokyo: Springer) pp 337–54

[132] Simoes-Torigoe R, Chen P-H, Li Y M, Kohanfars M, Morris K, Williamson C W, Makale M, Mayadev J and Talke F 2021 Design and validation of an automated dilator prototype for the treatment of radiation induced vaginal injury *2021 43rd Annual Int. Conf. of the IEEE Engineering in Medicine & Biology Society (EMBC)* (Piscataway, NJ: IEEE) pp 1562–5 doi: 10.1109/EMBC46164.2021.9630126

[133] Kitamori H, Sumida I, Tsujimoto T, Shimamoto H, Murakami S and Ohki M 2019 Evaluation of mouthpiece fixation devices for head and neck radiotherapy patients fabricated in polyjet photopolymer by a 3D printer *Phys. Med.* **58** 90–8

[134] Hong C-S, Oh D, Ju S G, Ahn Y C, Na C H, Kwon D Y and Kim C C 2019 Development of a semi-customized tongue displacement device using a 3D printer for head and neck IMRT *Radiat. Oncol.* **14** 1–8

[135] Mundee T, Jongwannasiri C and Fuangrod T 2022 Design of 3D-printed universal oral stent for tongue immobilization in head and neck radiotherapy *Biomed. Phys. Eng. Express* **9** 015011

[136] Kairn T, Chan P, Chua B, Cleland S, Dawes J, Kenny L, Lin C Y, McDowall W R, Poroa T and Crowe S 2023 Magnetic resonance imaging assessment of the suitability and consistency of radiotherapy treatment positioning achieved using intra-oral stents *IFMBE Proc.* (in press)

[137] Cleland S *et al* 2022 Development of a customisable 3D-printed intra-oral stent for head-and-neck radiotherapy *Tech. Innov. Patient Support Radiat. Oncol.* **23** 1–7

[138] Tanooka M, Ishida T, Moridera K, Ichimiya K, Tarutani K, Kitajima K, Fujiwara M, Kishimoto H and Kamikonya N 2017 Utility of intraoral stents in external beam radiotherapy for head and neck cancer *Rep. Pract. Oncol. Radiother* **22** 310–8

[139] Stieb S *et al* 2020 The impact of tongue-deviating and tongue-depressing oral stents on long-term radiation-associated symptoms in oropharyngeal cancer survivors *Clin. Transl. Radiat. Oncol.* **24** 71–8

[140] Singh A, Rosen E B, Randazzo J D, Estilo C L, Gelblum D Y and Huryn J M 2021 Intraoral radiation stents–primer for clinical use in head and neck cancer therapy *Head Neck* **43** 4010–7

[141] Nilsson S, Moutrie Z, Cheuk R, Chan P, Lancaster C, Markwell T, Dawes J and Back P 2015 A unique approach to high-dose-rate vaginal mold brachytherapy of gynecologic malignancies *Brachytherapy* **14** 267–72

[142] Zhang G-H, Lv X-M, Wu W-J, Wu Z-Y, Zheng L, Huang M-W, Wang Y and Zhang J-G 2019 Evaluation of the accuracy of computer-assisted techniques in the interstitial brachytherapy of the deep regions of the head and neck *Brachytherapy* **18** 217–23

[143] Jones E-L, Baldion A T, Thomas C, Burrows T, Byrne N, Newton V and Aldridge S 2017 Introduction of novel 3D-printed superficial applicators for high-dose-rate skin brachytherapy *Brachytherapy* **16** 409–14

[144] Park S-Y, Kang S, Park J M, An H J, Oh D H and Kim J i 2019 Development and dosimetric assessment of a patient-specific elastic skin applicator for high-dose-rate brachytherapy *Brachytherapy* **18** 224–32

[145] D'Alimonte L, Ravi A, Helou J, Morrison D, Mendez L C, Easton H and Morton G 2019 Optimized penile surface mold brachytherapy using latest stereolithography techniques: a single-institution experience *Brachytherapy* **18** 348–52

[146] Douglass M J and Santos A M C 2019 Application of optical photogrammetry in radiation oncology: HDR surface mold brachytherapy *Brachytherapy* **18** 689–700

[147] Laan R C, Nout R A, Dankelman J and van de Berg N J 2019 MRI-driven design of customised 3D printed gynaecological brachytherapy applicators with curved needle channels *3D Print. Med.* **5** 1–8

[148] Jiang P, Qu A, Wei S, Sun H, Zhang X, Li X and Wang J 2020 The preliminary results of 3-dimensional printed individual template assisted 192Ir high-dose rate interstitial brachytherapy for central recurrent gynecologic cancer *Technol. Cancer Res. Treat.* **19** 1533033820971607

[149] Imber B S, Wolden S L, Stambuk H E, Matros E, Wexler L H, Drew A S, Rosen E B, Ganly I, Gil'ad N C and Damato A L 2019 Novel intraoperative radiotherapy utilizing prefabricated custom three-dimensionally printed high-dose-rate applicators *Brachytherapy* **18** 277–84

[150] Sekii S, Tsujino K, Kosaka K, Yamaguchi S, Kubota H, Matsumoto Y, Ota Y, Sasaki R and Soejima T 2018 Inversely designed, 3D-printed personalized template-guided interstitial brachytherapy for vaginal tumors *J. Contemp. Brachyther* **10** 470–7

[151] Yuan X, Zhang Y, Cui M, Miao J, Gao L, Hu J, Tian D and You J 2019 Dosimetry comparison between a 3D printed minimally invasive guidance template and free implantation in the brachytherapy treatment of postoperative recurrent cervical carcinoma *Cancer Manage. Res.* **11** 5013

[152] Liu J-Y, Man Q-W, Ma Y-Q and Liu B 2017 I125 brachytherapy guided by individual three-dimensional printed plates for recurrent ameloblastoma of the skull base *Br. J. Oral and Maxillofacial Surg.* **55** e38–40

[153] Shi Y, Song Y, Liu T-Z, Liu Z-W, Huang M-W and Zhang J-G 2016 The design and application of 3D printing individual noncoplanar templates guiding brachytherapy in oral and pharynx area *Brachytherapy* **15** S66

[154] Jiang Y *et al* 2018 Side effects of CT-guided implantation of 125I seeds for recurrent malignant tumors of the head and neck assisted by 3D printing non co-planar template *Radiat. Oncol.* **13** 1–7

[155] Chen P, Wu W-J, Yi Z-Q, Ma X-L, Zhao W-H and Zhang J-G 2019 125 125I interstitial brachytherapy in management of pediatric skull base tumors *Pediatric Blood Cancer* **66** e27622

[156] Subashi E, Jacobs C, Hood R, Kirsch D G and Craciunescu O 2020 A design process for a 3D printed patient-specific applicator for HDR brachytherapy of the orbit *3D Print. Med.* **6** 1–5

[157] Huang M-W, Zhang J-G, Zheng L, Liu S-M and Yu G-Y 2016 Accuracy evaluation of a 3D-printed individual template for needle guidance in head and neck brachytherapy *J. Radiat. Res.* **57** 662–7

[158] Ji Z, Jiang Y, Guo F, Sun H, Fan J, Zhang L and Wang J 2017 Dosimetry verification of radioactive seed implantation for malignant tumors assisted by 3D printing individual templates and CT guidance *Appl. Radiat. Isot.* **124** 68–74

[159] Poulin E, Gardi L, Fenster A, Pouliot J and Beaulieu L 2015 Towards real-time 3D ultrasound planning and personalized 3D printing for breast HDR brachytherapy treatment *Radiother. Oncol.* **114** 335–8

[160] Aristei C *et al* 2019 Individualized 3D-printed templates for high-dose-rate interstitial multicathether brachytherapy in patients with breast cancer *Brachytherapy* **18** 57–62

[161] Guinot J L *et al* 2018 GEC-ESTRO ACROP recommendations in skin brachytherapy *Radiother. Oncol.* **126** 377–85

[162] Stephens H, Deans C, Schlect D and Kairn T 2020 Development of a method for treating lower-eyelid carcinomas using superficial high dose rate brachytherapy *Phys. Eng. Sci. Med.* **43** 1317–25

[163] Harris B D, Nilsson S and Poole C M 2015 A feasibility study for using ABS plastic and a low-cost 3D printer for patient-specific brachytherapy mould design *Australas. Phys. Eng. Sci. Med.* **38** 399–412

[164] Bieleda G, Chiche A, Boehlke M, Zwierzchowski G, Chyrek A, Burchardt W, Stefaniak P, Wiœniewska N, Czereba K and Malicki J 2022 3D printing of individual skin brachytherapy applicator: design, manufacturing, and early clinical results *J. Contemp. Brachyther* **14** 205–14

[165] Arenas M, Sabater S, Sintas A, Argus M, Hernández V, Árquez M, López I, Rovirosa À and Puig D 2017 Individualized 3D scanning and printing for non-melanoma skin cancer brachythcrapy: a financial study for its integration into clinical workflow *J. Contemp. Brachyther* **9** 270–6

[166] Bassi S, Langan B and Malone C 2019 Dosimetry assessment of patient-specific 3D printable materials for HDR surface brachytherapy *Phys. Med.* **67** 166–75

[167] Gonzalez-Perez V *et al* 2022 H&n and Skin (HNS) gec-estro working group critical review of recommendations regarding prescription depth, bolus thickness and maximum dose in skin superficial brachytherapy with flaps and customized moulds *Radiother. Oncol.* **175** 122–32

[168] Ricke J, Mohnike K, Pech M, Seidensticker M, Rühl R, Wieners G, Gaffke G, Kropf S, Felix R and Wust P 2010 Local response and impact on survival after local ablation of liver metastases from colorectal carcinoma by computed tomography-guided high-dose-rate brachytherapy *Int. J. Radiat. Oncol. Biol. Phys.* **78** 479–85

[169] Chitti B, Goyal S, Sherman J H, Caputy A, Sarfaraz M, Cifter G, Aghdam H and Rao Y J 2020 The role of brachytherapy in the management of brain metastases: a systematic review *J. Contemp. Brachyther* **12** 67–83

[170] Sapienza L G *et al* 2019 Detection of air gaps around the cylinder by postinsertion computed tomography in vaginal cuff brachytherapy: a prospective series, systematic review, and meta-analysis *Brachytherapy* **18** 620–6

[171] Magné N, Chargari C, SanFilippo N, Messai T, Gerbaulet A and Haie-Meder C 2010 Technical aspects and perspectives of the vaginal mold applicator for brachytherapy of gynecologic malignancies *Brachytherapy* **9** 274–7
[172] Crowe S B, Pross Z, Nilsson S, Dawes J, Kairn T and Lancaster C M 2018 Characterisation of radiological properties of a brachytherapy moulding material *Australas. Phys. Eng. Sci. Med.* **41** 731–7
[173] Wilks R, Crowe S B, Chan P, Cheuk R and Kairn T 2023 A 3D printed patient-specific vaginal mould for brachytherapy *J. Phys.: Conf. Ser.* **2630** 012026
[174] Zhao Z, Tang X, Mao Z and Zhao H 2019 The design of an individualized cylindrical vaginal applicator with oblique guide holes using 3D modeling and printing technologies *J. Contemp. Brachyther* **11** 479–87
[175] Rivard M J, Coursey B M, DeWerd L A, Hanson W F, Saiful Huq M, Ibbott G S, Mitch M G, Nath R and Williamson J F 2004 Update of AAPM Task Group no. 43 report: a revised AAPM protocol for brachytherapy dose calculations *Med. Phys.* **31** 633–74
[176] Logar H B Z, Hudej R and Šegedin B 2019 Development and assessment of 3D-printed individual applicators in gynecological MRI-guided brachytherapy *J. Contemp. Brachyther* **11** 128–36
[177] Beaulieu L 2012 Report of the Task Group 186 on model-based dose calculation methods in brachytherapy beyond the TG-43 formalism: current status and recommendations for clinical implementation *Med. Phys.* **39** 6208–36
[178] Lee J H, Kim H N, Lim H S and Cho S O 2019 Three-dimensional-printed vaginal applicators for electronic brachytherapy of endometrial cancers *Med. Phys.* **46** 448–55

IOP Publishing

3D Printing in Radiation Therapy

Tanya Kairn, Scott B Crowe and Tomas Kron

Chapter 10

Treatment verification

Deepak Basaula, Emily Simpson-Page, Scott B Crowe and Tanya Kairn

10.1 Introduction

In its broadest sense, radiation therapy treatment verification encompasses a range of medical physics tasks, from point dose and calibration measurements, to quality control of individual treatment plans, to the many measurements involved in treatment modality commissioning, and to the measurements needed for the research into new treatment techniques and technologies that make quality improvement possible. Three-dimensional (3D)-printed items can contribute significantly in all of these areas.

This chapter discusses some of the many opportunities for 3D printing to assist in the performance of verification measurements for radiation therapy systems and patient treatments. Specifically, this chapter addresses the applications of 3D printing for the design and fabrication of: bespoke jigs, attachments, and adaptors to aid in brachytherapy source strength calibration and external beam radiation therapy (EBRT) linac calibration (Sections 10.2.1 and 10.2.2); geometrically simple phantoms for use in imaging studies or dosimetry studies (section 10.3.2 and 10.3.3); and sophisticated anthropomorphic phantoms for use in research studies or treatment modality commissioning work (section 10.4).

The ability to create bespoke devices using a 3D printer is one of the key attractions for implementing 3D printing in a radiation therapy centre. The ability to customise both the design and fabrication aspects provides limitless opportunities for tailor-made solutions. The in-house fabrication of phantoms, jigs, and attachments is also the area where the monetary cost savings associated with 3D printing are most obvious (see chapter 6).

Various phantoms are used in radiation therapy to verify the safe and accurate performance of treatment equipment and ensure that patient treatments are closely matched to their prescriptions. Imaging phantoms are used to verify the performance of imaging systems that are used for treatment planning and patient setup, and dosimetry phantoms are used to measure radiation dose. Accordingly, imaging phantoms generally contain structures with clearly defined geometries, markers, and other features that can be used to assess specific properties of the imaging systems, whereas dosimetry phantoms are generally characterised by their radiological tissue-similarity and their ability to contain radiation dosimeters at one or more defined locations. Liquid water or water-equivalent plastics are commonly used as dosimetry phantom materials because they provide a homogeneous approximation of human tissue, although many situations (including radiation therapy system commissioning, new treatment technique commissioning, end-to-end testing, and radiation therapy research) require the use of more complicated and heterogeneous phantoms. Commercial solutions in this area are generally very costly and generic enough to be suited to a wide range of purposes with minimal adaptability.

The introduction of a 3D-printing service in a radiation therapy centre can open up a new horizon in designing and creating phantoms and other radiation therapy verification equipment, due to the ability to fabricate simple and complicated devices that are designed according to local clinical needs.

10.2 Dosimeter augmentation

10.2.1 Dosimetry jigs

An appropriately commissioned 3D-printing system (see chapter 8) should be able to produce 3D-printed objects with sufficient precision (or with appropriate geometric margins) to produce objects that function as reasonably complex dosimetry jigs, with interlocking parts.

For example, Simpson-Page *et al* [1] fabricated a brachytherapy source strength (reference air kerma rate) measurement jig in eight interlocking parts, including interchangeable attachments to allow the brachytherapy source to be held at multiple distances from two different models of Farmer-type ionisation chamber (see figure 10.1). Printing in sections (as shown in 10.1(a)) allowed the brachytherapy calibration jig to be flat-packed for mailing to 24 different radiation therapy centres in Australia and New Zealand for the purpose of completing a brachytherapy dosimetry audit [2]. The design of this jig was created using open source TinkerCAD software (see chapter 4) with reference to publicly available Farmer chamber geometries [1]. The reproducible and robust nature of a jig like this can be a cost-effective solution for a radiation therapy centre wishing to verify brachytherapy source strength with a Farmer chamber.

Wilks *et al* [3] described the production and use of a 3D-printed jig to hold a brachytherapy applicator in a standard 3D-scanning water tank. This jig, which was again designed using TinkerCAD, attached stably to the side of the water tank and allowed the applicator to be suspended in water (or air) 20 cm from the tank wall [4]. Placing a small-volume ionisation chamber in a holder attached to the water tank's built-in scanning arm made it possible to measure dose profiles and isodose distributions in water and in air, at a range of distances from a high dose-rate brachytherapy source, for the purpose of verifying calculations from a brachytherapy model-based dose calculation algorithm (MBDCA) [3] (see chapter 9).

Obviously, 3D printing allows similar jigs to be made for verification measurements of EBRT systems. Simple adaptors and attachments for this purpose are described in section 10.2.2.

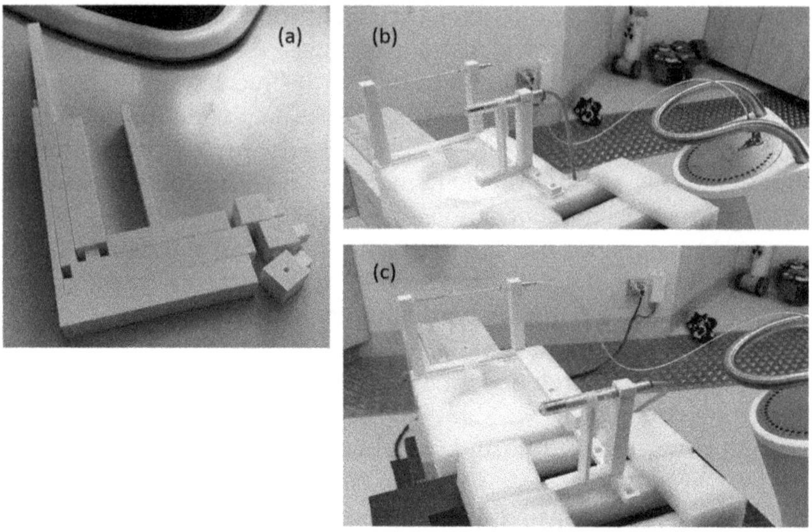

Figure 10.1. Photographs of the 3D-printed brachytherapy calibration jig described in [1], showing (a) 3D-printed parts, (b) jig in use with chamber at standard calibration distance and (c) jig in use, showing range of chamber positioning options (holes in horizontal strut).

When designing a jig, simplicity is important to ensure structural integrity and ease of use. Jigs should be designed so that they cannot be used incorrectly, and consequently any connections should be unique [5]. Jigs should be thoroughly commissioned before clinical implementation, especially if they are being used in situations where the jig may affect the measurement results, either by radiation scatter or attenuation. If this is the case, then the effect will need to be accurately characterised [6].

10.2.2 Adaptors, attachments, and inserts

The capabilities provided by a well-managed clinical 3D-printing services have given individual radiation therapy centres the ability to design and fabricate adaptors, attachments, and inserts that can extend the uses (and useful lifetimes) of existing dosimetry equipment. The ability to use 3D printing to replace parts that are broken, missing, or no longer in production [7, 8] is especially valuable during periods of supply-chain interruption [9]. Additionally, the ability to create attachments that extend the range of measurements that can be completed using existing dosimetry equipment has economic and also dosimetric benefits for healthcare providers.

Aftab *et al* [10] used 3D-printed attachments to interface with pre-existing ionisation chamber holders to investigate the dosimetric impact of a metallic drive rail in their 3D-scanning water tank. The use of 3D-printed attachments allowed chambers to be placed at various distances from the rail, to evaluate its potential effect on photon and electron beam dosimetry [10]. Perrett *et al* [11] proposed the use of 3D-printed caps, to allow diodes to be used for small field dosimetry in water phantoms without the need for density corrections derived from alternative measurements or sophisticated Monte Carlo calculations [12, 13].

Several authors have used 3D-printed inserts to interface with solid dosimetry phantoms, including film inserts to extend the range of the locations, sizes, and shapes of lung tumours for which dose measurements can be made in lung phantoms [14, 15] (see figure 10.2(a)), as well as 3D-printed blocks containing tumours and fiducial markers, to replace standard heterogeneous inserts for use with array dosimeters (see figure 10.2(b)). (More sophisticated methods for combining 3D-printed anthropomorphic phantom components with commercial array dosimeters are discussed in section 10.4.)

When designing a new phantom or jig with a cavity for a chamber, it is necessary to assess the fit of the cavity. However, this fitting process, which may involve iterative adjustments, could potentially be damaging to fragile and costly dosimeters. For this purpose, it can be beneficial to have a 3D model library of dosimeters of varying dimensions ready for printing and use in situations like this (figure 10.3).

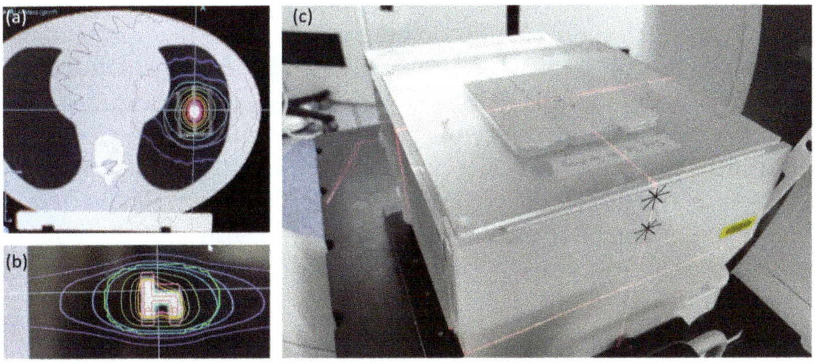

Figure 10.2. Examples of inserts for existing dosimetry phantoms: (a) and (b) 3D-printed film inserts for commercial moving phantom, including complex 'tumour' geometries. (b) 3D-printed lung blocks (central white slabs at the intersection of the lasers) containing spherical tumour and fiducial markers inserted into a standard heterogeneous phantom accessory of a commercial ionisation chamber array.

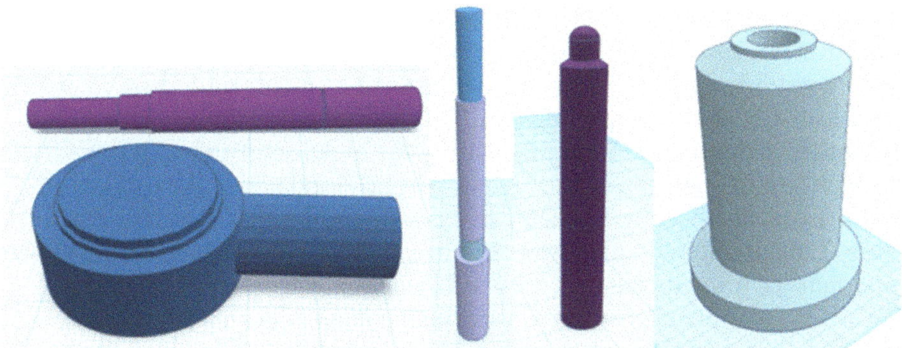

Figure 10.3. Example library of virtual dosimeter designs.

Dimensions of dosimetry equipment can be measured directly or obtained from dosimeter manufacturers. These model dosimeters can then be used for all testing, fitting, and other situations where the use of the real dosimeters is inadvisable.

When 3D printing a small part to fix an existing piece of equipment, it is important to ensure that the printed part is structurally sound. This may involve stress testing or reviewing specifications from the original manufacturer to establish that 3D printing a small attachment or accessory and replacing it on an existing piece of equipment will not reduce the integrity of the system. Though it may be tempting to overlook commissioning when printing such small parts, it is important to ensure the printed design meets specifications before it is used clinically or for any treatment verification purpose.

10.3 Geometrically simple phantoms

10.3.1 Hidden targets

Hidden-target testing is an important tool for the verification of image-guided radiation therapy (IGRT), which becomes more important with the increasing adoption of hypo-fractionated and stereotactic radiation therapy techniques that require especially precise patient positioning for every treatment fraction [16–19]. Before IGRT treatments can be delivered with a linac, it is necessary to locate that linac's radiation isocentre and confirm a close match with the location of the linac's imaging isocentre so that setup shifts derived from images of patients treated with that linac will correctly relate to the radiation isocentre. Radiation isocentre localisation is a fundamental linac quality assurance test, and the coincidence of the associated imaging isocentre is very important for the delivery of accurate and safe patient treatments.

The Winston–Lutz test is the workhorse of linac treatment-imaging isocentre coincidence verification (it has been long been used—before the advent of IGRT—for treatment isocentre localisation [16]) and it can be completed using a geometrically simple phantom. The key requirement of a Winston–Lutz test phantom is that it contains a small and regular target with a density that makes it visible in images from both the treatment beam and the IGRT imaging beam, and that it be rigidly attached to the linac's treatment couch via a system that allows fine scale adjustment of the phantom's position. This small phantom is then placed at the megavoltage (MV) treatment isocentre (located using images from the MV treatment beam, with multiple rotations of the gantry, collimator, and couch) and used to verify the centring of the IGRT imaging isocentre, or vice versa. Ideally, a flat surface on the outside of the phantom with lines indicating the location of the target inside can be used to confirm the position of the linac's light field and adjust the room lasers, after the MV treatment isocentre has been located. Figure 10.4(a) shows an example of a Winston–Lutz phantom, which was 3D printed from polylactic acid (PLA) with an internal marker printed using a copper-PLA composite material [17].

The advent of more sophisticated treatment couches—which allow translations in three dimensions, as well as pitch, roll, and yaw rotations—and more dosimetrically challenging treatments—which can include treatments of multiple, small, off-axis

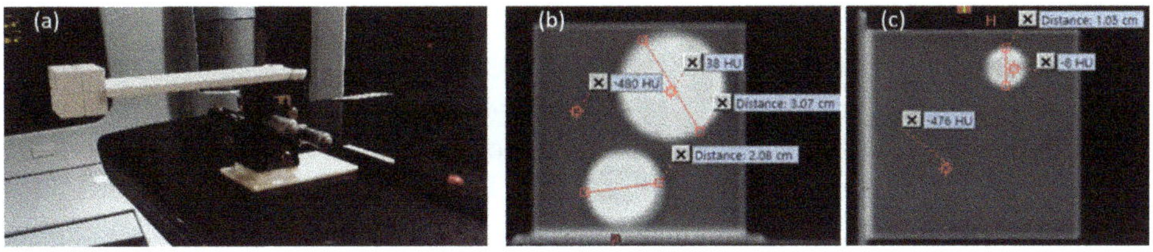

Figure 10.4. (a) Photograph of 3D-printed Winston–Lutz isocentre phantom. (b) and (c) images of 3D-printed multi-target cube.

targets treated using volumetric modulated arc therapy (VMAT) beams centred on one point—has led to a need for more sophisticated hidden-target phantoms to verify these new features. For example, figures 10.4(b) and (c) show images of a 3D-printed multi-target cube with different sized internal spheres offset from the centre to test couch translation accuracy.

Woods *et al* [18] designed and 3D printed a modified Winston–Lutz phantom where the usual one central ball-bearing marker was enclosed in a phantom with uneven sides and uneven external markings, which was used for daily matching with a computed tomography (CT) image where the phantom was scanned while sitting level. The purpose of this test was to establish that the radiation therapy treatment couch would drive the phantom to its correct level position, applying translations and rotations as necessary [18].

For routine geometric verification of multi-target single-isocentre treatment capability, Oliver *et al* [19] 3D printed a test phantom from high-impact polystyrene into which five tungsten carbide ball bearings were inserted, and then planned a treatment that was routinely used to acquire MV images showing the centring of each ball bearing in its corresponding treatment aperture for a range of gantry, collimator, and couch positions. This test effectively verified the IGRT system, which was used for initial phantom setup, as well as the reliability of the gantry, collimator, couch, and beam collimation system for accurately treating small and off-axis targets [19].

Mertens *et al* [20] further extended the range of tests that could be completed using a Winston–Lutz phantom by 3D printing a phantom from PA12, which included surface features for testing an optical surface guidance system and holes for a tissue insert, a lung insert and dosimeter for VMAT end-to-end testing, and also the marker for the fundamental Winston–Lutz test.

As EBRT treatment equipment becomes more sophisticated, with a growing range of patient imaging, positioning, and treatment delivery capabilities, the benefits of having access to an in-house programme for 3D printing purpose-designed devices to test this equipment becomes increasingly clear.

10.3.2 Simple imaging phantoms

When using imaging for radiation therapy, the three key areas that require verification, to ensure that patients are treated safely and accurately are image quality, geometric accuracy, and density reliability. Verification tests in all three areas can be enabled or assisted through the use of 3D-printed phantoms.

Image quality is tied to radiation therapy treatment effectiveness because images are often used to determine the extent and locations of the cancer, as well as the surrounding organs at risk, and so poor-quality images may lead to treatments that are ineffective or dangerous. There are numerous commercial phantoms that have been designed to assess image quality for CT and magnetic resonance imaging (MRI) systems, many of which are packaged (with associated analysis software) as part of the purchase of new imaging systems. By contrast, commercially-available phantoms for evaluating positron emission tomography (PET) image quality (signal detectability) are more limited, largely due to features of the technology itself. PET image quality can depend on the particular tracer chosen, its dosage and resulting uptake in the imaged tissues, the size of each target, and the thickness of tissue surrounding each target, and consequently different image quality phantoms are needed for different purposes [21]. These PET phantoms can be expensive and difficult to customise [22] and lack realism due to the presence of solid walls around each tracer receptacle [23].

The potential for 3D printing to contribute in the area of novel PET image quality phantom fabrication was demonstrated in early work by Bieniosek *et al* [22], who benchmarked their 3D-printing process by duplicating an existing commercial phantom and then used their MJP process to 3D print a novel phantom to verify image resolution for an ultra-high resolution PET scanner. To avoid the problem of solid walls around tracer receptacles, while including a realistic background signal, Wollenweber *et al* [23] used 4–10 mm diameter polyhedron wireframes printed in Nylon (PA) as ‘features’ in their phantoms filled with small acrylic spheres, so that the tracer would fill the polyhedrons entirely while minimally penetrating between the surrounding spheres. More recently, Läppchen *et al* [24] have investigated the addition of radioisotopes to liquid 3D-printing resins and Cerviño *et al* [25] have 3D printed a sophisticated cylindrical insert for a commercial respiratory motion phantom, so that PET images acquired with controlled motion could be evaluated.

Geometric accuracy is important for radiation therapy imaging due to its obvious impact on whether the radiation therapy treatment can hit its target. Any sufficiently large phantom that has radio-opaque features with known dimensions can be used for geometric verification of radiological images, including radiation therapy planning CT images and treatment setup images. The situation becomes more complex when dealing with MRI and ultrasound, two modalities that are important for radiation therapy but notorious for susceptibility to geometric deformations.

Although common 3D-printing materials are generally unsuitable for use in ultrasound phantoms (which need flexible materials that are speed-of-sound matched to human tissues), Füzesi and Gyöngy [26] provided an example where filament arrays for ultrasound resolution testing were 3D printed from acrylonitrile butadiene styrene (ABS), PLA, and a digital light processing photopolymer and were then immersed in water for imaging. 3D printing also has an obvious application in the area of fabricating bespoke moulds for gels and other mouldable materials that are suitable for ultrasound imaging (see chapter 3).

MRI is widely used to delineate tumour volumes due to the excellent soft-tissue contrast that is achievable with this modality. Consequently, geometric distortion of

MRI scans is concerning due to the potential consequences of delineating a target in an incorrect location or with an incorrect shape. Appropriate MRI verification phantoms are therefore needed to quantify and make allowances (e.g. margins) for geometric distortions. The potential for 3D printing to assist in this area has been demonstrated by Ramachandran *et al* [27], who used a grid phantom that was printed from PA12 using an selective layer sintering (SLS) process and then filled with water to assess differences in MRI image distortion with and without the application of a distortion correction, and to thereby show that 3D-printed phantoms such as this could be a cost-effective solution for radiation therapy centres wanting to use MRI for treatment planning and IGRT.

While 3D-printed components have been included in much more sophisticated anthropomorphic MRI phantoms [28], for the purpose of MRI-based IGRT in particular, Powers *et al* [29] used simple 3D-printed PLA shells to contain MRI-visible fluid as part of their commissioning of an MR-linac system. The resulting phantom contained features that were visible in MRI and CT images, and were therefore suitable for geometric verification and end-to-end testing of the MR-linac. These phantom components are shown in figure 10.5.

The density and relative electron density (RED) data derived from CT images used for radiation therapy treatment planning needs to be accurate and consistent due to its role in the calculation of radiation dose. In radiation therapy treatment planning, the Hounsfield unit value in each voxel of the 3D planning CT image is converted into a corresponding RED value (and sometimes a material definition, depending on the dose calculation algorithm) to be used to calculate the 3D treatment dose distribution using a previously measured CT-RED calibration table [30]. Measurement of this CT-RED calibration data, and subsequent periodic verification of its continued reliability is achieved using a purpose-designed CT-RED phantom.

Several commercial models of CT-RED phantoms are available and have been established for providing reliable RED data for treatment planning when used

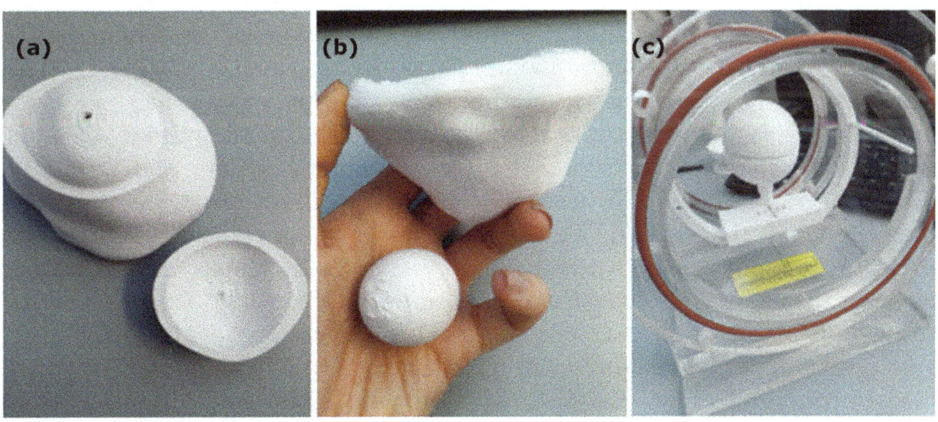

Figure 10.5. A simple MRI phantom for IGRT geometry testing: (a) Prostate shape with small hole for filling with MRI-visible liquid, (b) anatomical and spherical designs, (c) stage for supporting 3D-printed components inside external housing to be filled with MRI-visible liquid.

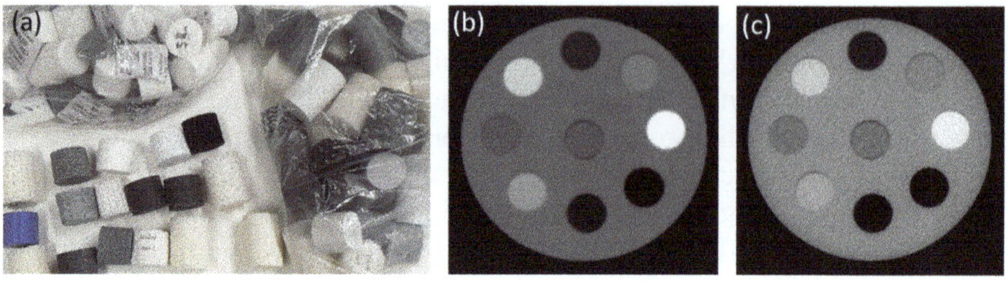

Figure 10.6. (a) Photograph of 3D-printed CT-RED phantom insert plugs of varying densities. (b) kV CT and (c) MV CT images of CT-RED calibration phantom containing plugs of varying densities.

appropriately [31, 32]. However, 3D-printing offers the ability to create additional inserts for these phantoms to add extra detail to calibrations or to investigate the properties of specific materials. Examples of this type of work are shown in figure 10.6 [14, 33].

An additional possible role for 3D printing in the production of geometrically simple phantoms for imaging purposes is in the area of radiomics [34]. For example, Varghese *et al* [35] created a 3D-printed texture phantom comprising a six-texture pattern, and used it to evaluate the robustness and reproducibility of a radiomics panel under a routine abdominal imaging protocol.

10.3.3 Simple dosimetry phantoms

The published use of 3D-printed geometrically simple phantoms is comparatively rare given that the major advantage of 3D printing in the treatment verification space lies in the ability to design and create complex geometries including patient-matched anthropomorphic phantoms (see section 10.4). However, there are situations where simple phantoms are needed and yet commercial phantoms are still too generic for the purpose. Figure 10.7 shows 3D-printed dosimetry phantoms that were designed for investigations of dosimetrically challenging scenarios. In each of these cases, the geometry and/or density of an anatomical structure has been abstracted to allow measurements of particular parameter without including extraneous variables by being too complicated or too patient-specific.

Figures 10.7(a) and (b) show a 3D-printed rectilinear phantom that was used to compare dose calculations from a treatment planning system and an adaptive radiation therapy system [36] without needing to extract dose points along the curved surfaces that would have existed in a more-realistic geometry [37]. Figures 10.7(c) and (d) show an eyeball that was designed using averaged geometric measurements from MRI images and 3D printed at a constant density to allow an investigation of the differences between calibration factors identified for a Sr-90 pterygia treatment plate arising from measurements in this phantom compared to measurements made using a flat surface [38]. Figures 10.7(e) and (f) show a 3D print of a breast containing a void designed to accommodate an intra-operative radiation therapy applicator. In this case, the heterogeneous tissues (which would include

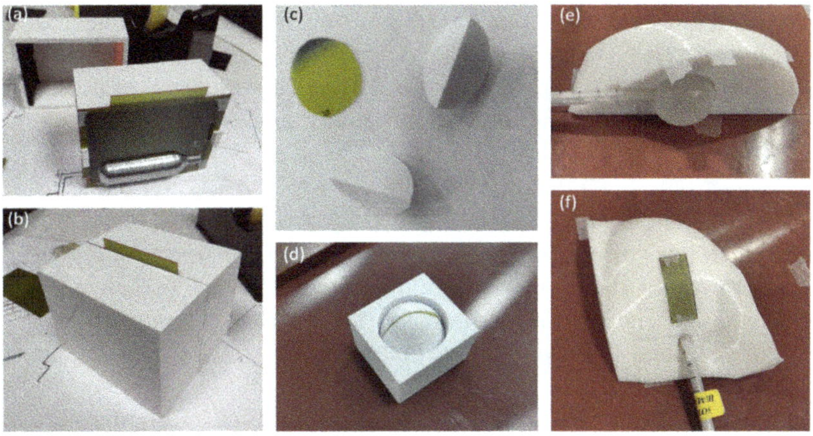

Figure 10.7. Simple 3D-printed dosimetry phantoms (a) and (b) abstracted breast phantom containing internal air volume and CO_2 canister for modelling gas-filled temporary tissue expander; (c) and (d) average eyeball phantom for measuring dose from Sr-90 plate; and (e) and (f) a homogeneous breast phantom for measuring dose from intra-operative radiation therapy treatment.

muscle, glandular, and adipose soft tissues) were homogenised to focus the investigation on the effects of source positioning and calibration.

General designs for these abstracted phantoms arise from the process of identifying the dosimetry question that needs to be answered, isolating the variables that are important to the question, and simplifying away the variables that are not relevant. Specific design details may require additional planning and measurements, and particular care should be taken to ensure that the chosen simplifications are not themselves introducing bias. For example, in designing the phantom shown in figures 10.7(e) and (f), it was necessary to create additional 3D-printed sheets of the proposed phantom material and use them in a basic transmission study using a superficial kilovoltage (kV) EBRT beam to establish the limitations and uncertainties affecting the eventual kV intra-operative radiation therapy measurements.

Relatedly, a key form of simple dosimetry phantom is the 3D-printed sheet phantom, which needs to be 3D printed in a range of thicknesses and then used for transmission (or percentage depth dose or tissue-maximum ratio) measurements as part of the process of planning, designing, and creating an anthropomorphic phantom (see section 10.4, also see chapter 8). The key requirement of these simple prints is that the properties of the sheets must be as similar as possible to the properties of the materials when they are included in the anthropomorphic phantoms, so that observations made using the sheets will remain approximately valid when considering the usually larger and more complicated phantoms.

10.4 Anthropomorphic phantoms

10.4.1 Summary of requirements

Conventional imaging and dosimetry phantoms are specifically designed for the structured testing of specific aspects of system performance (e.g. the high contrast resolution of an imaging system using a phantom containing specific geometric and radiological properties). While these phantoms allow efficient, reliable, and quantifiable tests of one or more steps of a diagnostic or treatment process, they are often unsuitable for the testing of all components under real-world conditions because of the degree of abstraction of clinical reality.

The testing of whole diagnostic or treatment processes under clinical conditions, or end-to-end testing, is widely recommended by professional bodies [39, 40]. It may be performed as part of an independent audit of a local practice; for example, for Level III dosimetric audits of radiation therapy centres [41] and clinical trial accreditations [42]. The phantoms used for these tests are often anthropomorphic (i.e. they are designed to have human characteristics). The characteristics can be broadly categorised as:

- Geometric, that is the dimensions of the phantom and components approximate anatomical dimensions;
- Material, that is the phantom material approximates the radiological, magnetic resonance, acoustic, or other physical properties of tissue; and/or
- Functional, that is the phantom can replicate the behaviour of human anatomy such as motion or deformation.

The degree of similarity to a human or human body part will depend on the intended application, and all phantom designs require some level of approximation. For example, the most widely-used thorax phantom for radiation therapy (CIRS phantom Model 002LFC, Computerised Imaging Reference Systems Inc, Norfolk, USA), which is recommended for treatment planning system commissioning by the International Atomic Energy Agency and widely used for audits and end-to-end testing [43, 44], is elliptical in shape, with discrete uniform soft-tissue, lung densities, a cylindrical bone-equivalent spine, and is incapable of motion or deformation. The CIRS ATOM phantoms are more sophisticated generic phantoms that are designed to mimic the external shape of a typical person (male or female, paediatric to adult), with internal lung- and bone-equivalent structures. Kyoto Kagaku's anthropomorphic phantoms are similarly designed as typical person models (also male or female, paediatric to adult), including posable limbs and joints, for training and testing of x-ray imaging systems (Kyoto Kagaku, Kyoto, Japan). A more complicated version of the CIRS thorax phantom (CIRS Model 008A) contains bones modelled on the general shape of human vertebra rather than being a simple cylinder and is capable of programmable motion [45, 46]. Phantoms like these often require some degree of modularity to allow a range of tests to be performed; for example, commercial radiation therapy phantoms frequently include cavities to allow insertion of dosimeters or contrast materials [46].

The use of patient imaging data to design and manufacture phantoms through 3D printing is gaining significant popularity [47, 48]. 3D-printed phantoms can be cheaper than commercially-available phantoms (see chapter 6) and can be designed to test processes for which no commercial alternatives exist. 3D-printing technology allows the production of anthropomorphic phantoms that replicate selected geometric, material, and functional properties of humans, as described in the following sections (section 10.4.2, 10.4.3 and 10.4.4).

10.4.2 Geometric properties

Geometric anthropomorphism can be achieved by basing the phantom dimensions on designs approximating human anatomy; for example, developed within computer aided design software, the use of existing virtual or physical phantoms, or the use of acquired or synthesised medical imaging data. Software that can be used for phantom design and to convert medical images to 3D model files is described in chapter 4.

Medical imaging data is commonly used to design anthropomorphic phantoms, as described in chapter 4. Anthropomorphic phantoms may take the form of specific organs (e.g. liver, lung, etc), or body parts (e.g. head, hand, etc), or anatomical regions (e.g. thorax, pelvis, etc) or specific sections of anatomical regions (e.g. transverse slice of brain, sagittal slice of thorax, etc) and may be used for many different applications (e.g. dosimetry, image reconstruction and registration, deformation, etc) [47, 49–53]. The use of clinical medical imaging data for such purposes requires ethical review, but there are several open access repositories of medical imaging data for which conditional approval for education or research use may be already granted. For example, the Cancer Imaging Archive [54] is a considerable repository of medical images that may be accessible for conversion to 3D-printed phantoms. Software used to produce synthetic medical imaging data or virtual phantoms [48], such as XCAT, can also be used to design phantoms for printing [55]. In addition, several authors have augmented or reproduced commercial anthropomorphic phantoms with 3D printing [14, 56, 57]. For example, Tino *et al* [14] created a new 3D-printed section for an existing anthropomorphic phantom by CT scanning one of the phantom's slices, segmenting the different tissue types, adding holes for dosimeters, and then adding a previously segmented lung tumour to one of the parts, as shown in figure 10.8.

The impact of print errors such as warping can scale with the phantom's dimensions, which may be large for anthropomorphic phantoms. Craft and Howell [58] described the printing of a sagittal-sliced thorax phantom where warping caused air gaps between printed slices. Experimentation may be necessary to minimise these effects (e.g. by modifying orientation of the print on the print bed, the use of platform additions such as brims or rafts, or the optimisation of temperatures, as described in chapters 5 and 8).

10.4.3 Material properties

Chapter 3 describes various 3D-printing materials that can be used as surrogates for human tissue. Phantom materials can be directly printed using 3D-printing techniques, or indirectly moulded using a printed cavity or shell.

The behaviour of ionising radiation within a material is described by physical cross sections that depend on the radiation type and energy, and the composition and structure of the material, or more specifically the mass, electron, proton, and neutron densities. By adjusting these parameters, the attenuation and dosimetric properties of a phantom can be altered for both diagnostic and therapeutic radiation qualities.

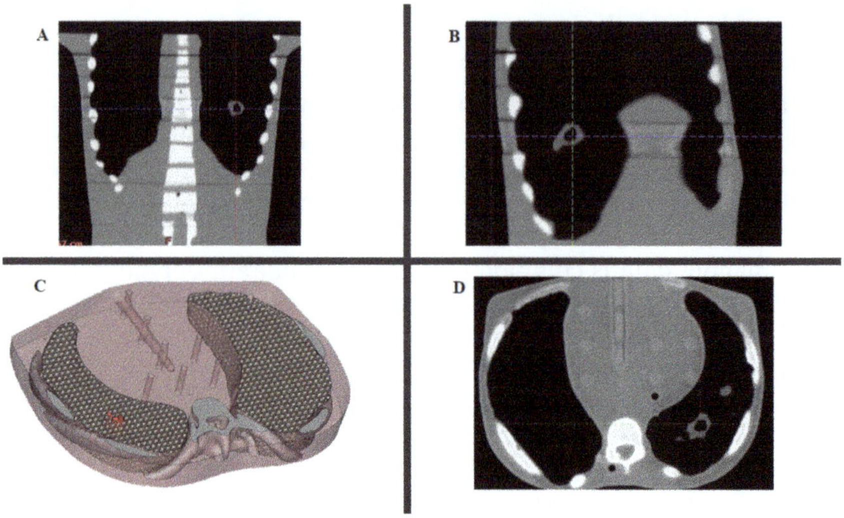

Figure 10.8. Illustrations of a 3D-printed phantom segment integrated with a commercial phantom, as described in [14]: (A) coronal view, (B) sagittal view, (C) 3D view of a 3D-printed segment showing lung- and bone-equivalent regions newly added lung tumour and holes for dosimeter placement, and (D) transverse view.

Materials with densities lower than commonly printed plastics (i.e. < 1.1 g cm^{-3}), such as lung tissue, can be approximated by variation of infill density. For low infill densities, an isotropic infill pattern such as a gyroid pattern should be used to reduce any directional dependence effects in the attenuation properties of the material [59].

Materials with higher densities than common plastics, such as cortical bone, can be mimicked through the use of composites containing higher density materials, such as stone powders [55, 56, 60] or metals [33, 61], or by the addition of materials after printing, such as gypsum powder [62]. The use of surrogate materials containing high concentrations of elements or molecules dissimilar to those found in the body (e.g. metal composites) will exhibit varying behaviour for different radiation qualities [63] (see chapter 8).

Material and infill densities in a phantom can typically be specified on a per-model basis in most printer slicing software packages (as described in chapter 5), which can be achieved through threshold segmentation of medical imaging data based on pixel values and production of multiple STLs [64, 65], or direct voxel-by-voxel printing [66].

The use of a printer supporting more than one material, such as a dual-nozzle printer [17, 56, 61] or supplementary devices such as the Mosaic Palette [67], can allow the fabrication of soft tissues with both variable infill densities and bone.

The relationship between nominal material density, infill density, and attenuation properties will vary with filament and printer parameters [68], and should be characterised by the reader for their own radiation therapy centre for the desired energies using imaging or transmission measurements [6, 65, 69–73] or composition analysis [74]. For phantoms used for radiation dosimetry, care must be taken to

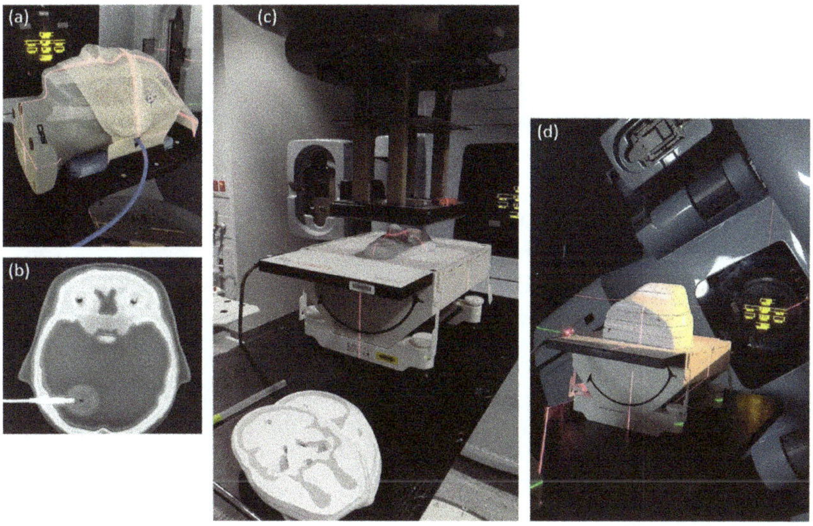

Figure 10.9. Examples of anthropomorphic phantoms in use for dosimetry purposes: (a) and (b) 3D-printed transverse slice added to a commercial head phantom to allow end-to-end testing with an ionisation chamber, (c) coronally sliced 3D-printed head phantom in use for electron radiation therapy verification using an array dosimeter, (d) sagittally-sliced 3D-printed thorax phantom in use for hypo-fractionated lung radiation therapy treatment verification.

apply appropriate correction factors to measurements based on the radiation type and scatter conditions [74–76].

Figure 10.9 shows examples of multi-material phantoms and phantom sections that were 3D printed using dual-nozzle systems from filaments that were previously established to be suitable for replicating the effects of bone, lung, and tissue in kV and MV treatment beams [56]. Figures 10.9(a) and (b) show an example resulting from combining the dosimeter models discussed in section 10.2.1 with a CT scan of an existing commercial phantom to produce a 3D-printed phantom slice that fits the existing phantom, including a CT-visible tumour and an ionisation chamber cavity for dose measurements, similar to the work described by Tino *et al* [14]. Figures 10.9(c) and (d) show 3D-printed phantoms in use with a standard array dosimeter (Octavius 729 ionisation chamber array, PTW Dosimetry, Freiburg, Germany) for the purpose of radiation therapy treatment dose verification. The phantom design used in figure 10.9(c) allows the array to be placed at various depths downstream of the heterogeneous and irregular 3D-printed face for end-to-end testing of electron radiation therapy beams in a challenging but clinically-realistic geometry; there was no commercially-available phantoms suitable for this purpose. The sagittally-sliced phantom in figure 10.9(d) allows the array to be placed at the depth of a peripheral lung tumour, as well as at out-of-field positions to assess dose to healthy lung and the heart, for the purpose of assessing important dosimetric features of a new hypo-fractionated lung radiation therapy technique [55].

The three examples shown in figure 10.9 highlight three key areas where 3D-printed anthropomorphic phantoms have the potential to make a substantial contribution to radiation therapy verification:

- Figures 10.9(a) and (b) show a situation where multiple commercial solutions exist but are much more complicated and expensive than the 3D-printed solution that was used;
- Figure 10.9(c) shows a common treatment verification scenario for which no suitable commercial phantom currently exists; and
- Figure 10.9(d) shows a 3D-printed phantom that was created to investigate and verify a potential new treatment technique.

10.4.4 Functional properties

While there has been increasing interest in cardiac motion (relating to cardiac radio-ablation [77]) and peristalsis (relating to oesophageal radiation therapy [78]), and there has long been interest in bowel gas motion (relating to prostate radiation therapy [79]), there is no doubt that the main biological function of concern for radiation therapy treatment planning and delivery is respiratory motion [46].

Currently, radiation therapy phantom motion can be achieved using commercial motion platforms [80], commercial dynamic phantoms [45, 46], custom solutions incorporating rotary and linear actuators [81], and even robotic arms [82]. The potential opportunities to contribute to this area through 3D printing include the design and fabrication of inserts for existing moving phantoms (see section 10.2.2) that may use more clinically-realistic tumour geometries or variable lung densities, as well as the design and use of more comprehensive systems, where 3D printing is used in the drive system to achieve the particular motion needed [83], as well as for the fabrication of the phantom itself.

Deformable phantoms have been produced by 3D-printing moulds to cast deformable materials such as silicone [84, 85], softened PVC [86, 87], and gels [88]. For example, Matrosic *et al* [88] used 3D printing to mould gel dosimeters that were then embedded in a deformable phantom to investigate the effects of respiratory motion on radiation therapy dose in the abdominal region and confirm the dosimetric advantages of treatment beam gating as a respiratory motion management strategy.

Deformable phantoms have also been created by directly 3D printing the deformable anatomy, via careful selection of materials and printing methods. For example, Shin *et al* [89] used a rubberlike material and an MJP technique to 3D print a flexible bronchial structure, which was then embedded in a cylinder of flexible polyurethane foam and driven by a moving diaphragm-like platform, and evaluated using repeated four-dimensional (4D) CT imaging. As a more complete example, Colvill *et al* [28] created an MRI and CT compatible deformable thorax phantom with multiple organs that were moulded or directly 3D printed depending on their required deformation characteristics.

Phantoms such as these, which include interchangeable 3D-printed components, have potential uses in deformable registration studies, 4D CT and 4D MRI evaluations, and investigations of particularly challenging clinical cases for patients with non-standard anatomy. As radiation therapy treatment techniques become more sophisticated and motion management options become both more numerous and more imperative, the need for sophisticated moving and deformable phantoms will only increase. 3D printing clearly has an important role to play in this area.

10.5 Key points

- 3D-printed jigs, attachments, and adaptors can be used to augment existing dosimeters to extend the breadth and longevity of the dosimeters' usability.
- Simple imaging phantoms can be 3D printed for use in verifying the three key features of radiation therapy images: image quality (especially for PET systems), geometric accuracy (especially for ultrasound and MRI systems), and density reliability (especially for CT systems).
- In-house fabrication of simple dosimetry phantoms allows verification measurements to be performed for challenging clinical scenarios with reference to the key variables that impact dosimetry in those scenarios.
- Detailed anthropomorphic phantoms can be 3D printed to excellent effect if care and attention is given to the geometric, material, and functional requirements of the phantoms.

References

[1] Simpson-Page E, Hamlett L, Lew D, Stephens H, Wilks R, Kairn T and Crowe S B 2021 3d printed brachytherapy jig for Reference Air Kerma Rate calibration *Phys. Eng. Sci. Med.* **44** 1141–50

[2] Simpson-Page E, Wilks R, Kairn T and Crowe S 2023 Postal audit for all high dose-rate brachytherapy centres in Australia and New Zealand: lessons learned *Phys. Eng. Sci. Med.* **46** 469

[3] Wilks R, Peet S, Kairn T and Crowe S 2023 Adapting a 3d scanning water phantom for use in brachytherapy dosimetry *IFMBE Proc.* (in press)

[4] Wilks R, Skewes J, Forrestal D, Kairn T and Crowe S 2023 Evaluating 3D printing materials for intracavitary gynaecological brachytherapy *Phys. Eng. Sci. Med.* **46** 482

[5] Brans K 2013 3D printing, a maturing technology *IFAC Proc.* **46** 468–72

[6] Madamesila J, McGeachy P, Barajas J E V and Khan R 2016 Characterizing 3D printing in the fabrication of variable density phantoms for quality assurance of radiotherapy *Phys. Med.* **32** 242–7

[7] Livingstone A G 2022 3d printed PTW TRUFIX-compatible detector holders *Phys. Eng. Sci. Med.* **45** 356

[8] Sim L 2016 Novel application of 3D printing in brachytherapy using MED610 3D printed insert for I-125 ROPES eye plaque *Australas. Phys. Eng. Sci. Med.* **39** 863–70

[9] Nazir A, Azhar A, Nazir U, Liu Y-F, Qureshi W S, Chen J-E and Alanazi E 2021 The rise of 3D printing entangled with smart computer aided design during covid-19 era *J. Manuf. Syst.* **60** 774–86

[10] Aftab S, Barnes M P, Doebrich M and Lehmann J 2020 The effect of the horizontal metallic drive on reference dosimetry in the SNC 3D scanner water tank *J. Appl. Clin. Med. Phys.* **21** 95–101

[11] Perrett B, Charles P, Markwell T, Kairn T and Crowe S 2017 Feasibility of 3D printed air slab diode caps for small field dosimetry *Australas. Phys. Eng. Sci. Med.* **40** 631–42
[12] Charles P, Cranmer-Sargison G, Thwaites D, Kairn T, Crowe S, Pedrazzini G, Aland T, Kenny J, Langton C and Trapp J 2014 Design and experimental testing of air slab caps which convert commercial electron diodes into dual purpose, correction-free diodes for small field dosimetry *Med. Phys.* **41** 101701
[13] Cranmer-Sargison G, Weston S, Evans J, Sidhu N and Thwaites D 2011 Implementing a newly proposed Monte Carlo based small field dosimetry formalism for a comprehensive set of diode detectors *Med. Phys.* **38** 6592–602
[14] Tino R B, Yeo A U, Brandt M, Leary M and Kron T 2022 A customizable anthropomorphic phantom for dosimetric verification of 3D-printed lung, tissue, and bone density materials *Med. Phys.* **49** 52–69
[15] Kairn T and Crowe S 2019 Producing inserts for a commercial motion phantom using 3D printing *Australas. Phys. Eng. Sci. Med.* **42** 377–8
[16] Rowshanfarzad P, Sabet M, O'Connor D J and Greer P B 2011 Isocenter verification for linac-based stereotactic radiation therapy: review of principles and techniques *J. Appl. Clin. Med. Phys.* **12** 185–95
[17] Peet S C, Simpson-Page E, Livingstone A G and Crowe S B 2023 3D printing of high-density phantom materials for linear accelerator isocentre verification *IFMBE Proc.* (in press)
[18] Woods K, Ayan A S, Woollard J and Gupta N 2018 Quality assurance for a six degrees-of-freedom table using a 3D printed phantom *J. Appl. Clin. Med. Phys.* **19** 115–24
[19] Oliver P, Wood T and Baldwin L 2022 A customizable, open-source Winston–Lutz system for multi-target, single isocentre radiotherapy *Biomed. Phys. Eng. Express* **8** 065005
[20] Mertens L, Fleckenstein J, Steil V and Schneider F 2022 A novel end-to-end test for combined dosimetric and geometric treatment verification using a 3D-printed phantom *Med. Dosim.* **47** 177–83
[21] Mu'minah I and Soejoko D *et al* 2019 Phantom design for quality control of positron emission tomography/computed tomography (PET/CT) imaging *J. Phys.: Conf. Ser.* **1248** 012031
[22] Bieniosek M F, Lee B J and Levin C S 2015 Characterization of custom 3D printed multimodality imaging phantoms *Med. Phys.* **42** 5913–8
[23] Wollenweber S D, Alessio A M and Kinahan P E 2016 A phantom design for assessment of detectability in PET imaging *Med. Phys.* **43** 5051–62
[24] Läppchen T, Meier L P, Fürstner M, Prenosil G A, Krause T, Rominger A, Klaeser B and Hentschel M 2020 3d printing of radioactive phantoms for nuclear medicine imaging *EJNMMI Phys.* **7** 1–13
[25] Cerviño L *et al* 2017 A novel 3d-printed phantom insert for 4D PET/CT imaging and simultaneous integrated boost radiotherapy *Med. Phys.* **44** 5467–74
[26] Füzesi K and Gyöngy M 2017 Comparison of two inexpensive rapid prototyping methods for manufacturing filament target ultrasound phantoms *Ultrasound Med. Biol.* **43** 712–20
[27] Ramachandran P, Noble C, Langton C, Perrett B, Cox J, Chapman M and McGill G 2021 A 3D printed phantom to assess MRI geometric distortion *Biomed. Phys. Eng. Express* **7** 035004
[28] Colvill E *et al* 2020 Anthropomorphic phantom for deformable lung and liver CT and MR imaging for radiotherapy *Phys. Med. Biol.* **65** 07NT02

[29] Powers M *et al* 2022 Commissioning measurements on an Elekta Unity MR-linac *Phys. Eng. Sci. Med.* **45** 457–73

[30] Khan F M and Gibbons J P 2014 *Khan's the Physics of Radiation Therapy* (Baltimore, MD: Williams & Wilkins)

[31] Inness E K, Moutrie V and Charles P H 2014 The dependence of computed tomography number to relative electron density conversion on phantom geometry and its impact on planned dose *Australas. Phys. Eng. Sci. Med.* **37** 385–91

[32] Jarema T and Aland T 2019 Using the iterative kV CBCT reconstruction on the Varian Halcyon linear accelerator for radiation therapy planning for pelvis patients *Phys. Med.* **68** 112–6

[33] Tino R, Yeo A, Brandt M, Leary M and Kron T 2021 The interlace deposition method of bone equivalent material extrusion 3D printing for imaging in radiotherapy *Mater. Des.* **199** 109439

[34] Albarakati H, Jackson P, Geso M, Tino R and Kron T 2016 Radiomics feature stability on 4D CT- Evaluation with 3D printed textures in motion phantom *Phys. Eng. Sci. Med.* **44** 963–4

[35] Varghese B A, Hwang D, Cen S Y, Lei X, Levy J, Desai B, Goodenough D J and Duddalwar V A 2021 Identification of robust and reproducible CT-texture metrics using a customized 3D-printed texture phantom *J. Appl. Clin. Med. Phys.* **22** 98–107

[36] Kairn T, Lathouras M, Sylvander S, Trapp J and Crowe S 2022 Adaptive radiotherapy dosimetry in a challenging geometry: A model gas-filled tissue expander in a helical tomotherapy beam *J. Phys.: Conf. Ser.* **2167** 012012

[37] Kairn T, Lathouras M, Grogan M, Green B, Sylvander S R and Crowe S B 2021 Effects of gas-filled temporary breast tissue expanders on radiation dose from modulated rotational photon beams *Med. Dosim.* **46** 13–20

[38] Maxwell S, Crowe S, Simpson-Page E, Cassim N and Kairn T 2022 RSC: a 3D printed eyeball phantom for Sr-90 dosimetry measurements *J. Phys.: Conf. Ser.* **2167** 012018

[39] Smilowitz J B *et al* 2015 AAPM medical physics practice guideline 5.a.: commissioning and QA of treatment planning dose calculations—megavoltage photon and electron beams *J. Appl. Clin. Med. Phys.* **16** 14–34

[40] Geurts M W, Jacqmin D J, Jones L E, Kry S F, Mihailidis D N, Ohrt J D, Ritter T, Smilowitz J B and Wingreen N E 2022 AAPM medical physics practice guideline 5.b: commissioning and QA of treatment planning dose calculations–megavoltage photon and electron beams *J. Appl. Clin. Med. Phys.* **23** e13641

[41] Kron T, Hamilton C, Roff M and Denham J 2002 Dosimetric intercomparison for two Australasian clinical trials using an anthropomorphic phantom *Int. J. Radiat. Oncol. Biol. Phys.* **52** 566–79

[42] Clark C H, Hurkmans C W and Kry S F 2017 The role of dosimetry audit in lung SBRT multi-centre clinical trials *Phys. Med.* **44** 171–6

[43] Schiefer H *et al* 2010 The Swiss IMRT dosimetry intercomparison using a thorax phantom *Med. Phys.* **37** 4424–31

[44] IAEA Dosimetry and Medical Radiation Physics section 2016 *Commissioning of Radiotherapy Treatment Planning Systems: Testing for Typical External Beam Treatment Techniques* (Vienna: International Atomic Energy Agency)

[45] Shimohigashi Y, Yotsuji Y, Maruyama M, Kai Y and Toya R 2019 Target volume and motion position evaluation of four-dimensional cone-beam CT: comparison with 4D-CT using dynamic thorax phantom *Biomed. Phys. Eng. Express* **5** 035016

[46] Tajik M, Akhlaqi M M and Gholami S 2022 Advances in anthropomorphic thorax phantoms for radiotherapy: a review *Biomed. Phys. Eng. Express* **8** 052001

[47] Tino R, Yeo A, Leary M, Brandt M and Kron T 2019 A systematic review on 3D-printed imaging and dosimetry phantoms in radiation therapy *Technol. Cancer Res. Treat.* **18** 1533033819870208

[48] Xu X G 2014 An exponential growth of computational phantom research in radiation protection, imaging, and radiotherapy: a review of the fifty-year history *Phys. Med. Biol.* **59** R233

[49] Tino R, Leary M, Yeo A, Kyriakou E, Kron T and Brandt M 2020 Additive manufacturing in radiation oncology: a review of clinical practice, emerging trends and research opportunities *Int. J. Extreme Manuf.* **2** 012003

[50] McGarry C K *et al* 2020 Tissue mimicking materials for imaging and therapy phantoms: a review *Phys. Med. Biol.* **65** 23TR01

[51] Filippou V and Tsoumpas C 2018 Recent advances on the development of phantoms using 3D printing for imaging with CT, MRI, PET, SPECT, and ultrasound *Med. Phys.* **45** e740–60

[52] Mitsouras D *et al* 2015 Medical 3D printing for the radiologist *Radiographics* **35** 1965–88

[53] Mitsouras D, Liacouras P C, Wake N and Rybicki F J 2020 Radiographics update: medical 3D printing for the radiologist *Radiographics* **40** E21–3

[54] Prior F, Smith K, Sharma A, Kirby J, Tarbox L, Clark K, Bennett W, Nolan T and Freymann J 2017 The public cancer radiology imaging collections of The Cancer Imaging Archive *Sci. Data* **4** 1–7

[55] Kairn T, Jessen L, Bodnar J, Charles P H and Crowe S B 2023 Lung radiotherapy verification with a 3D printed thorax phantom and an ionisation chamber array *J. Phys.: Conf. Ser.* **2630** 012028

[56] Kairn T, Zahrani M, Cassim N, Livingstone A, Charles P and Crowe S 2020 Quasi-simultaneous 3D printing of muscle-, lung-and bone-equivalent media: a proof-of-concept study *Phys. Eng. Sci. Med.* **43** 701–10

[57] Ehler E D, Barney B M, Higgins P D and Dusenbery K E 2014 Patient specific 3D printed phantom for IMRT quality assurance *Phys. Med. Biol.* **59** 5763

[58] Craft D F and Howell R M 2017 Preparation and fabrication of a full-scale, sagittal-sliced, 3D-printed, patient-specific radiotherapy phantom *J. Appl. Clin. Med. Phys.* **18** 285–92

[59] Tino R, Leary M, Yeo A, Brandt M and Kron T 2019 Gyroid structures for 3D-printed heterogeneous radiotherapy phantoms *Phys. Med. Biol.* **64** 21NT05

[60] Okkalidis N and Marinakis G 2020 Accurate replication of soft and bone tissues with 3D printing *Med. Phys.* **47** 2206–11

[61] Mille M M, Griffin K T, Maass-Moreno R and Lee C 2020 Fabrication of a pediatric torso phantom with multiple tissues represented using a dual nozzle thermoplastic 3D printer *J. Appl. Clin. Med. Phys.* **21** 226–36

[62] Niebuhr N I, Johnen W, Güldaglar T, Runz A, Echner G, Mann P, Möhler C, Pfaffenberger A, Jäkel O and Greilich S 2016 Technical note: radiological properties of tissue surrogates used in a multimodality deformable pelvic phantom for MR-guided radiotherapy *Med. Phys.* **43** 908–16

[63] Crowe S B, Bennett J, Lathouras M, Lancaster C M, Sylvander S R, Chua B, Bettington C S, Lin C Y and Kairn T 2020 Impact of radiopacified bone cement on radiotherapy dose calculation *Phys. Imaging Radiat. Oncol.* **14** 12–6

[64] Nowak L J and Pawlowska E 2019 An algorithm and software for conversion of radiotherapy contour-sequence data to ready-to-print 3D structures *Med. Phys.* **46** 1829–32

[65] Leary M, Kron T, Keller C, Franich R, Lonski P, Subic A and Brandt M 2015 Additive manufacture of custom radiation dosimetry phantoms: an automated method compatible with commercial polymer 3d printers *Mater. Des.* **86** 487–99

[66] Okkalidis N 2018 A novel 3D printing method for accurate anatomy replication in patient-specific phantoms *Med. Phys.* **45** 4600–6

[67] Savi M, Andrade M A B and Potiens M P A 2020 Commercial filament testing for use in 3D printed phantoms *Radiat. Phys. Chem.* **174** 108906

[68] Craft D F, Kry S F, Balter P, Salehpour M, Woodward W and Howell R M 2018 Material matters: analysis of density uncertainty in 3D printing and its consequences for radiation oncology *Med. Phys.* **45** 1614–21

[69] Kairn T, Crowe S and Markwell T 2015 Use of 3D printed materials as tissue-equivalent phantoms *World Congress on Medical Physics and Biomedical Engineering, June 7–12, 2015, Toronto, Canada* (Berlin: Springer) pp 728–31

[70] Dancewicz O, Sylvander S, Markwell T, Crowe S and Trapp J 2017 Radiological properties of 3D printed materials in kilovoltage and megavoltage photon beams *Phys. Med.* **38** 111–8

[71] Michiels S *et al* 2016 Towards 3D printed multifunctional immobilization for proton therapy: initial materials characterization *Med. Phys.* **43** 5392–402

[72] Burleson S, Baker J, Hsia A T and Xu Z 2015 Use of 3D printers to create a patient-specific 3D bolus for external beam therapy *J. Appl. Clin. Med. Phys.* **16** 166–78

[73] Villani D, Rodrigues O Jr and Campos L 2020 Dosimetric characterization of 3D printed phantoms at different infill percentages for diagnostic x-ray energy range *Radiat. Phys. Chem.* **172** 108728

[74] Charles P H, Crowe S and Kairn T 2021 Recommendations for simulating and measuring with biofabricated lung equivalent materials based on atomic composition analysis *Phys. Eng. Sci. Med.* **44** 331–5

[75] Charles P H, Crowe S B and Kairn T 2021 Small field dose correction factors for radiochromic film in lung phantoms *Med. Phys.* **48** 2667–72

[76] Diamantopoulos S, Kantemiris I, Patatoukas G, Dilvoi M, Efstathopoulos E, Kouloulias V and Platoni K 2018 Theoretical and experimental determination of scaling factors in electron dosimetry for 3d-printed polylactic acid *Med. Phys.* **45** 1708–14

[77] Lydiard P, Blanck O, Hugo G, O'Brien R and Keall P 2021 A review of cardiac radioablation (CR) for arrhythmias: procedures, technology, and future opportunities *Int. J. Radiat. Oncol. Biol. Phys.* **109** 783–800 Suzanne

[78] Mostafaei F, Tai A, Omari E, Song Y, Christian J, Paulson E, Hall W, Erickson B and Li X A 2018 Variations of MRI-assessed peristaltic motions during radiation therapy *PLoS One* **13** e0205917

[79] Crook J, Raymond Y, Salhani D, Yang H and Esche B 1995 Prostate motion during standard radiotherapy as assessed by fiducial markers *Radiother. Oncol.* **37** 35–42

[80] Kairn T, Mitchell J, Crowe S and Trapp J 2015 Interplay of MLC, gantry and respiratory motion during DCAT delivery *World Congress on Medical Physics and Biomedical Engineering, June 7–12, 2015, Toronto, Canada* (Berlin: Springer) pp 428–31

[81] Pallotta S, Calusi S, Foggi L, Lisci R, Masi L, Marrazzo L, Talamonti C, Livi L and Simontacchi G 2018 ADAM: a breathing phantom for lung SBRT quality assurance *Phys. Med.* **49** 147–55

[82] Capaldi D P, Skinner L B, Dubrowski P, Zhang H, Xing L, Chuang C F, Loo B W, Bush K K, Fahimian B P and Amy S Y 2021 A robotically assisted 3D printed quality assurance lung phantom for Calypso *Phys. Med. Biol.* **66** 074005

[83] Ahmad M A, Mustaza S M, Mokri S S, Azmi N A, Ahmad R, Ramli R, Rahman W N W A and Abd Rahni A A 2020 Development of a 3D printed motion mechanism for a 4D respiratory motion phantom *2020 IEEE-EMBS Conf. on Biomedical Engineering and Sciences (IECBES)* (Piscataway, NJ: IEEE) pp 132–5

[84] Kadoya N *et al* 2017 Evaluation of deformable image registration between external beam radiotherapy and HDR brachytherapy for cervical cancer with a 3d-printed deformable pelvis phantom *Med. Phys.* **44** 1445–55

[85] Ehrbar S *et al* 2019 ELPHA: dynamically deformable liver phantom for real-time motion-adaptive radiotherapy treatments *Med. Phys.* **46** 839–50

[86] Liao Y *et al* 2017 An anthropomorphic abdominal phantom for deformable image registration accuracy validation in adaptive radiation therapy *Med. Phys.* **44** 2369–78

[87] He Y *et al* 2019 Characterizing mechanical and medical imaging properties of polyvinyl chloride-based tissue-mimicking materials *J. Appl. Clin. Med. Phys.* **20** 176–83

[88] Matrosic C K, Hull J, Palmer B, Culberson W and Bednarz B 2019 Deformable abdominal phantom for the validation of real-time image guidance and deformable dose accumulation *J. Appl. Clin. Med. Phys.* **20** 122–33

[89] Shin D-S, Kang S-H, Kim K-H, Kim T-H, Kim D-S, Chung J-B, Lucero S A, Suh T S and Yamamoto T 2020 Development of a deformable lung phantom with 3D-printed flexible airways *Med. Phys.* **47** 898–908

IOP Publishing

3D Printing in Radiation Therapy

Tanya Kairn, Scott B Crowe and Tomas Kron

Chapter 11

Beyond radiation therapy

Mathilde R Desselle and Natalka Suchowerska

11.1 Introduction

The motivation to customise treatments and devices to individual patient needs is not unique to radiation therapy. When we consider each individual case by case, we tailor the medical response to support optimal function or recovery, enabling people to have the best possible quality of life. Such an individualised approach has always been expensive and time-consuming to achieve. Now three-dimensional (3D) printing provides an accessible solution to some of these challenges.

To date, the most popular and emerging uses of 3D printing in medicine have been in creating patient-matched anatomical models, constructing templates for medical interventions, building surgical guides, building customised prostheses and orthoses, regenerative medicine, and bioprinting.

Although it can be argued that these applications have limited relevance to 3D printing in radiation therapy, the fundamental concepts are not dissimilar. In the immediate term, surgical guides (section 11.4) are relevant for brachytherapy, while prostheses and regenerative medicine (sections 11.5 and 11.6) can be expected to play an increasing role in radiation therapy planning. Medical physicists employed in radiation therapy need to be ready when called upon to support technical developments in fields beyond radiation therapy. Some applications of 3D printing have been more successful than others and further applications are emerging. It is common to find materials that are typically used in industry for their material properties being considered (for example) to replace machined-titanium implants for cancer patients. Some applications of 3D printing in medicine beyond radiation therapy will be discussed in this chapter.

11.2 Patient-matched anatomical models

3D printing of phantoms for the purpose of patient-specific dosimetry or dosimetric audits has been described in chapter 10. The printing of anatomical models can also support surgical planning, patient information, education, and implant design. Each application demands subtly different criteria in their design, construction, and material choices, such that they warrant a separate discussion.

First, anatomical models for surgical planning are one of the most common 3D-printed devices currently used in medicine in a broad range of surgical subspecialties. They are extensively used in oro-maxillo-facial and musculoskeletal surgery [1]. The 3D-printed models allow for surgical teams to gain an understanding of the patient's anatomy and its spatial relationships, and can support procedure rehearsal and plate pre-bending. This practice can result in reduced operating time, and increased surgical accuracy and surgeon confidence, ultimately leading to improved post-operative outcomes, including reductions in blood loss and hospital-acquired infections [2].

3D printing in healthcare is a highly multidisciplinary endeavour, where each stakeholder has responsibility for a component of the digital workflow, as highlighted in figure 11.1. If a clinician envisages a patient-matched anatomical model, then the first step will be to consult with biomedical engineers and physicists to design a case plan. This will inform the acquisition of the required medical images by radiology, generally using computed tomography (CT) or magnetic resonance imaging (MRI) modalities, or multi-modal image fusion depending on the tissue types and the best compromise between image resolution and patient radiation exposure (similar to the recommendations for imaging radiation therapy devices provided in chapter 4). A high-resolution narrow slice thickness is preferred for 3D modelling. The resulting DICOM file is communicated to the 3D-printing team, who will process the images using specialist software, clean and delete any artifacts, and segment the image object to isolate the anatomy of interest. The segmented files are processed into a virtual 3D model for visualisation and reported to the treating clinician for approval. Once approved by the clinician, and if a physical model is required, the relevant combination of additive and subtractive manufacturing technologies and materials is employed to create a physical device and post-process

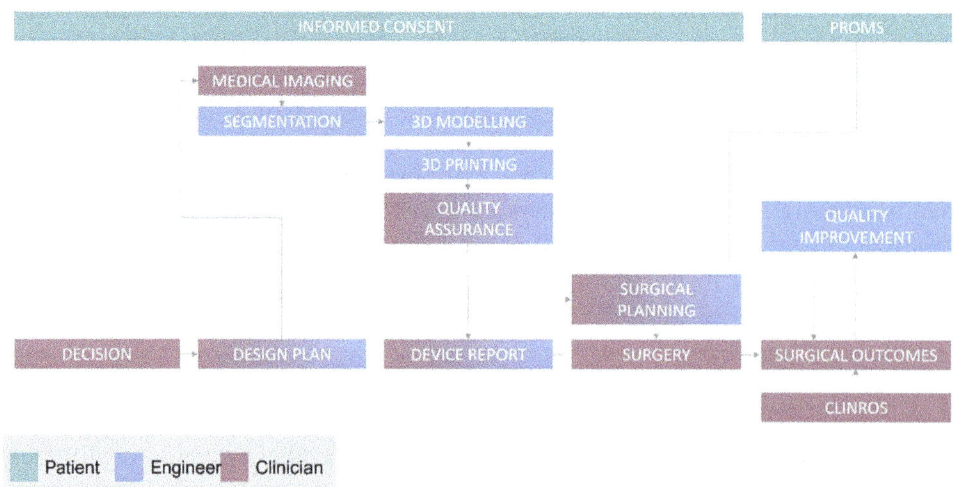

Figure 11.1. Simplified manufacturing workflow and stakeholder involvement for anatomical models for surgical planning.

to specifications. After the relevant quality assurance assessments and regulatory declarations, the device is then dispatched to the clinician to fulfil its intended purpose. Following surgery, clinician-reported outcomes measures (ClinROs) and patient-reported outcome measures (PROMs) should be employed to complement the evaluation of post-surgical outcomes, which in turn will inform quality improvement for engineering processes and future patients.

Numerous manufacturing options are available, depending on the requirements and intended use of the device (see the examples in chapter 2). Factors to be considered include resolution, speed of manufacture and post-processing, cost, the use of rigid or flexible materials, or a combination of both, colour, therapeutic radiation hardness and diagnostic radiation opacity, biocompatibility, and importantly any requirements for sterilising where heat treatment or radiation may compromise the geometric accuracy and physical integrity of the model. As a result, the actual cost of device production varies greatly according to the choice of material, software, hardware, infrastructure, and labour. Some of these steps in creating a 3D-printed object are fully integrated in the hospital system, while some can be outsourced depending on the nature of the device [3]. Readers may want to explore the detailed guide to image processing for the creation of 3D anatomical models for the planning of orthopaedic surgery published by Green *et al* [4].

Importantly, users of anatomical models for surgical planning must be familiar with the local regulatory framework for medical devices, which is evolving in many countries. In Australia, in 2021, the safety and performance of anatomical models for surgical planning came under the scrutiny of the Therapeutic Goods Administration, who are responsible for the regulation of medical devices in Australia. The reform saw the previously unregulated 3D-printed models jump up to a risk Class IIa medical device when used for surgical planning, resulting in new requirements for quality management processes associated with the production of patient-matched anatomical models. (Quality management for 3D printing is addressed in chapter 7.)

Creating these models is a multidisciplinary endeavour where surgeons and biomedical engineers and physicists work together to improve clinical effectiveness and demonstrate conformity to regulatory requirements. Irrespective of the specific steps in workflow conducted by clinicians, biomedical engineers, academics, or external contractors, the clinician is ultimately responsible for validating the accuracy of the design and model, and should therefore understand the materials, technical challenges, and regulatory requirements, and ensure that those involved in manufacturing hold the appropriate qualifications and comply with applicable standards. Murray *et al* [5] have interpreted the regulatory changes as of 2022 in a perspective article, to provide some guidance to surgeons.

Second, anatomical models can also be used as a tool to improve patient understanding in the provision of informed consent prior to a procedure. For example, in a case review, Chen *et al* [6] reported on a patient who, following a ureteric junction obstruction repair, did not return for the removal of the covering ureteric stent despite repeated correspondence from the treating team. Almost two years later, the patient presented to the emergency department with pain due to significant stent calcification but refused surgery to remove the stent. Using the patient's CT scan, the team modelled and 3D printed an anatomical model to help the patient understand his condition and the proposed procedure, resulting in the patient's eventual agreement to the surgery being performed [6].

Third, anatomical models are used for plate bending and implant design. Mathew *et al* [7] used an anatomical model to plan a tumour resection on the affected side of a patient's mandibula to complete the reconstruction based on the non-affected side of the mandibula, and to pre-bend a mandibular reconstruction plate and record screw placement. The establishment of this process for mandibula reconstruction suggests that 3D models can be used to design patient-matched implants that will replace the lost tissue in the context of other complex bone tumours or large segmental defects. As explained by Tetsworth *et al* [8], the design of a personalised implant is typically conducted by the team, which includes surgeons and biomedical engineers/physicists, in a virtual surgical planning session. The 3D model is used to discuss and simulate the reconstruction and to configure the implant. Creating a 3D model for virtual surgical planning and implant design has become standard practice in orthopaedic surgery. For example, Wong *et al* [9] performed a partial acetabular resection in a patient with a pelvic tumour, and used the anatomical model both to estimate resection margins under computer navigation guidance and to inform the design of a 3D-printed and patient-matched titanium implant.

While the processes to produce anatomical models may seem straightforward, the implications on practice are substantial. In considering the 3D printing of a personalised implant for a sarcoma patient, Randhawa *et al* [10] surveyed multiple stakeholders who each articulated different priorities: a nurse may be concerned by the anxiety caused to the patient by acquiring multiple radiology scans necessary to create the anatomical model and design the implant; a surgeon may be focused on the scans to ensure that the technology will improve surgical accuracy, but lacks the time to complete the paperwork necessary for regulatory compliance; and the biomedical engineers will need to understand the procedure, inform scan parameters, and efficiently communicate with the time-poor surgeon to review and approve designs (figure 11.2).

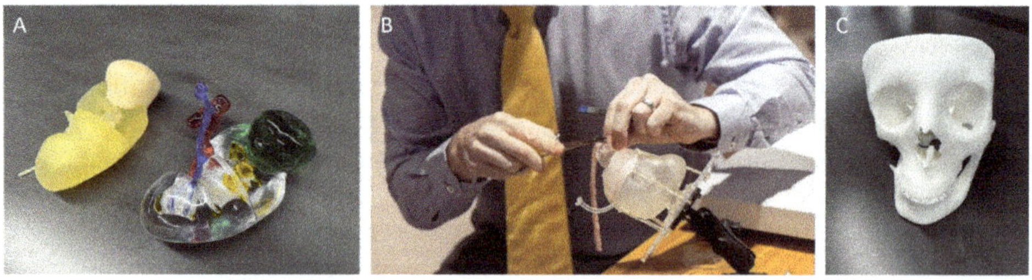

Figure 11.2. A. Kidney models for surgical planning 3D printing using polyjet technology. On the left-hand side, the model has been printed in a combination of flexible material (kidney) and hard material (tumour). On the right-hand side, the full colour model includes transparent materials to allow for visualisation of the tumour (green), vasculature (blue and red), and urine collection canal (yellow). **B.** A polyjet 3D-printed bladder model for surgical training of ureteral reimplantation. Courtesy of Nicholas Rukin and Rebecca Murray, Herston Biofabrication Institute/Redcliffe Hospital. C. A craniofacial reconstruction patient-specific model that is used for surgical planning and pre-bending of standard plates. Courtesy of Danilo Carluccio, Herston Biofabrication Institute/Royal Brisbane and Women's Hospital.

11.3 Templates for clinical intervention

The ability to visualise imaging data in 3D on a monitor has transformed surgery. We have seen how 3D printing enables faithful copies of the current patient anatomy and the proposed solution to be made, such that the surgeon is more confident, potentially having practiced the planned procedure and considered options to minimise risks. In cancer surgery, where no two cancers will be the same and discrimination of extent of cancerous tissue is not always obvious, surgical time can be reduced by an advance appreciation of the size and location of the tumour, relative to normal or sensitive structures and vasculature.

3D modelling and 3D printing afford for an infinite number of anatomies and pathologies to be reproduced, and for desired features to be designed to provide a range of options. 3D printing is extensively used to create simulation and training tools. This supports the practice of low frequency, low volume, complex, or high-risk but life-saving procedures without causing harm to critically ill patients.

The financial and ethical cost of access to animal and cadaver laboratories is prohibitive, and commercial surgical simulators are equally expensive. For example, access to a head and neck cadaver can cost around AU$3,000, and a 'fat neck' airway simulator around AU$5,000. Only one or two students can utilise the cadaver or simulator for a tracheostomy (creating an opening in the neck to insert a tube in the trachea to allow a trauma patient to breathe) before the cadaver is too damaged to be used again. 3D technologies can support the production of high-fidelity, synthetic, bespoke and patient-specific phantoms to simulate the characteristics of various tissue types, and the modification of commercial simulators. This approach also allows for modular surgical simulators, where only certain parts need to be replaced, and for practicing at different levels of difficulty (e.g. by recreating unusual anatomies). A common example is the 3D printing of flexible tracheas to replicate the qualities of standard ovine models while allowing training at different levels of complexity, and customising the airway to different sizes, anatomies, and pathologies (typical examples include obese neck and tracheal tumours and nodules) [11]. Another example is the use of 3D printing for the development of a low-cost simulation mannequin for extracorporeal cardiopulmonary resuscitation (ECPR), which is connected to a portable extracorporeal membrane oxygenation (ECMO) as a means of partial cardiopulmonary bypass. In an ECMO bypass, large tubes are inserted into a patient's blood vessels in the groin to provide oxygen to the blood. Intensivists Pang *et al* [12] modified a standard mannequin with a 3D-printed modular pelvis, and silicone and ballistic gel inserts to practice cannulation in a range of simulated anatomies. They developed an ECPR simulator for 10% of the cost of a commercial unit.

11.4 Surgical guides

Surgical guides are a complementary device to anatomical models for surgical planning. They provide a template for precise locations to cut and drill; for example, in the context of osteotomy, where the anatomical model is used to both to design the guide and to rehearse positioning it in the intended position. Surgical guides are extensively used in upper-limb, lower-limb, and maxillo-facial and dental surgery, and can lead to shorter operating room time and improved functional outcomes for the patient [13, 14].

The manufacturing of surgical guides for 3D surgical planning is more complex than that of anatomical models because of the attention that needs to be paid to the choice of materials. Surgical guides come in transient contact with the patient's body for cutting or drilling. As such, they must be produced in a material with a sufficient level of biocompatibility, which at the same time will maintain dimensional accuracy, physical properties, and biocompatibility after sterilisation. It is important that point-of-care manufacturing facilities, where steam sterilisation is likely to be used, should conduct their own part accuracy assessment to evaluate the effect of heat treatment on the materials [15] (also see chapter 9). Surgical guides must be made of material that can withstand the forces which may be applied to a template for drilling and cutting. In any case, a spare replica template should always be produced, sterilised, and ready for use in surgery.

11.5 Customised prostheses and orthoses

3D printing has been exploited to produce prosthetic limbs to replace missing anatomy or augment function. The traditional fabrication of prosthetics is labour-intensive, can be invasive for the patient, and is highly reliant on the skill and expertise of the practitioner creating the device. Traditional approaches involve manual plaster casting, moulding of thermoplastics, and cutting and shaping by an experienced practitioner. The use of digital workflows that rely on non-invasive surface 3D scanning and 3D printing to create prostheses has become popular, especially for children outgrowing their prosthesis. 3D-printed prostheses and orthoses can be customised to the patient's anatomy with a short turnaround and can be reprinted as often as needed. Once the primary print file has been established, any modifications can easily be incorporated. For example, the limb may be customised to look 'natural' or be adapted for a specific activity or sport. In this section, we explore the use of 3D printing for the creation of upper-limb, and facial prosthetics and orthoses.

The 3D printing of limb prosthetics affords customisation, not only for function and comfort but also for appearance. Paralympic athlete Anna Grimaldi received a 3D-printed titanium custom arm attachment for weightlifting. Teenage innovator Jordan Reeves became a worldwide sensation with Project Unicorn, when she designed a unicorn horn-shaped 3D-printed prosthetic arm that shoots glitter to advocate for those living with limb difference and for STEAM education [16]. Open-source 3D-printable designs of prosthetics created by independent DIY designers and tinkerers are extensively shared online, such as the 40,000 strong volunteers-led e-NABLE platform, which shares a curated library of upper-limb prosthetics designs that can be 3D printed using consumer-grade equipment and materials, as well as manufacturing and assembly resources [17]. However, without clinician input and acceptance, there can be a disconnect between the people developing such DIY devices and the needs of upper-limb prosthetics users. Wendo *et al* [18] recently reviewed the devices listed by the e-NABLE platform, as well as others, and concluded there is little literature documenting the clinical efficacy of such open-source devices. However, the sharing of open-source designs supports not only accessibility for people all around the world but also supports research and development efforts towards integration of upper-limb prosthetics with bionics and sensor systems [19].

Looking at lower-limb applications, 3D printing offers opportunities for more affordable and less time intensive manufacturing of customised prosthetic legs and legs sockets through scan to print approaches. The socket acts as a load transfer device between the residual limb and the prosthetic. 3D printing enables multiple iterations of sockets to be produced at an affordable cost, with the final design being wrapped in carbon fibre to ensure structural integrity [20].

Orthoses are another interesting application for 3D-printing. Non-invasive 3D surface scanning, and the ability to iterate designs and to create lightweight orthoses by adjusting 3D-printing techniques and materials, hold the potential to improve biomechanic outcomes and patient comfort with devices such as ankle-foot orthoses [21]. Another application of 3D printing for upper limbs is that of 3D-printed hand splints based on 3D surface scans in lieu of traditional plaster casts and thermoplastic splints. Again, there is little evidence of the superiority of such approaches in terms of functional outcomes. If patient-reported comfort is slightly higher with 3D-printed hand orthoses, then device fabrication using traditional approaches is considerably faster than with 3D-printing approaches because it can be completed during a patient's appointment [22]. However, like for hand splints, there is currently little evidence in the literature, prompting the need for further validation and clinical implementation research.

Ultimately, the adoption of 3D printing for the production of prostheses and orthoses is reliant on clinicians changing well-established, traditional manufacturing workflows to adopt digital scan to print workflows, which is an approach that is already well-established in dentistry for devices such as dental aligners [23]. The Royal Brisbane and Women's Hospital in Australia is currently exploring an occupational therapist-led approach for the production of transparent facial orthoses to assist in the management of scars in people who have experienced severe burns, which is already resulting in quicker production turnaround and improved patient experience, satisfaction, and comfort [24].

An existing widespread adoption of digital workflows for orthoses is the 3D printing of customised orthotic insoles, where the use of powder-based 3D-printing technologies leads to a reduction in production waste [25]. The use of 3D-printing moulds for the production of facial prosthetics, particularly ears and noses, is also well-documented [26]. Puls *et al* [23] recently reviewed the use of 3D printing for orbital and ocular prostheses, and found that 3D printing was mostly used to create moulds or parts for impressions, but as advanced manufacturing technologies evolve and more biocompatible materials become available, the direct 3D printing of orbital and ocular prostheses is becoming increasingly realistic (figure 11.3).

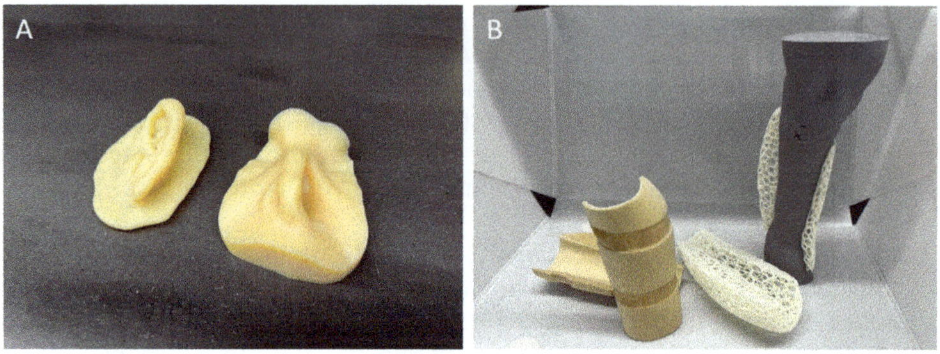

Figure 11.3. A. Facial prosthetic moulds created using 3D surface scanning and polyjet 3D printing. Courtesy of Danilo Carluccio, Herston Biofabrication Institute/Royal Brisbane and Women's Hospital. B. A foam lower-limb partial prosthetic (left-hand side) that caused discomfort to its user is replaced by a lightweight design that has been 3D printed using selective laser sintering. Courtesy of Roozbeh Fakhr, Herston Biofabrication Institute.

11.6 Regenerative medicine

Regenerative medicine has been described as the branch of medicine that develops methods to regrow, repair, or replace damaged or diseased cells, organs, or tissues. 3D printing has the ability to create the scaffold or hive, which the patient's own cells can colonise to create a permanent composite or a temporary composite material until the body recovers and regenerates. 3D printing has been successfully employed to support scaffold-guided bone tissue engineering, including in the context of large defects. The 3D-printed scaffold acts as a template for the patient's own tissue to grow and eventually integrate with or replace the implant.

Titanium 3D printing is used for the production of implants thanks to the material's biocompatibility and mechanical properties. Widespread applications include joints, spine, and extremities. Tetsworth *et al* [27] successfully used patient-matched 3D-printed titanium cages in multiple complex trauma patients with femoral bone loss. 3D printing also allows for enhancing implant performance through the customisation or modification of the surface geometry of the implant at the micro- or nano-level. This may lead to better surgical outcomes by improving osseointegration and cell adherence, or minimising the risk for infection [28].

Beyond titanium and titanium alloys, the scaffolds can also be 3D printed using bioresorbable polymers. For example, Kobbe *et al* [29] used a bioresorbable 3D-printed scaffold, combined with an autologous bone graft and growth factors, to repair an 11 cm long tibial defect. Here, 3D printing allows for the design of an osteoconductive scaffold, and the production of a pore geometry with the desired compromise between mechanical integrity and material degradation. Recently, comparable techniques have been employed for soft tissue. Cheng *et al* [30] successfully used a 3D-printed bioresorbable scaffold that was filled with an autologous fat graft to reconstruct a congenital chest wall deformity. These examples show the importance of collaboration between academia, hospitals, consumers, and industry. The same approach is currently being explored in a first-in-human clinical trial where 3D-printed breast scaffolds are used to replace silicone implants, anticipating that each patient's own tissue will have replaced the scaffold within 3 to 4 years [31]. Another emerging use is the seeding of 3D-printed bioresorbable scaffolds with cells to create personalised wound dressings, which may bring alternatives to skin grafting in the context of full thickness acute and chronic wounds [32]. Although the use of bioresorbable scaffolds for vascular stents has been attempted, a loss of structural integrity right after stent expansion was observed, leading to thrombosis [33].

Biocompatibility is not just about the material but also about using techniques that can impart biological function to the material of the implant. When the patient is young and otherwise healthy, the approach of using a resorbable material for 3D printing is desirable, avoiding the complications of living with a foreign body implant. However, for cancer patients the surgical loss of bone is often just the first step in a series of treatments, which are targeted at killing the patient's cancer cells and reducing opportunities for regrowth, which makes the resorbable implant approach less suitable. Solid commercial or 3D-printed titanium implants are common but for cancer patients requiring radiation therapy, titanium perturbs the radiation dose distribution, creating uncertainties and introducing special considerations for planning and treatment [34, 35]. Monitoring cancer progression in the presence of a titanium implant becomes radiologically impossible. Recently, Katsifis *et al* [36] have proposed an alternative with the use of polyether ether ketone (PEEK) polymers as a 3D-printable material, which behaves similarly to water in a therapeutic or diagnostic radiation beam. A composite design of PEEK and bone would optimise the implant performance [37]. Other materials in the PEEK family, such as polyetherketone (PEK) and polyetherketoneketone (PEKK), can be used as an alternative to titanium for non-load bearing cases.

A key disadvantage of PEEK is that it is bioinert. Although bone cells can colonise a PEEK implant, adhesion of regenerating bone tissue to the implant is critical to their long-term success. Careful attention needs to be paid to the surface roughness and surface energy [38]. The surface roughness can be modified either with skilled printing or with post-processing. An alternate approach, which was pioneered by Prof. David McKenzie at the University of Sydney, has been the use of plasma immersion ion implantation (PIII) to bury a layer of persistent carbon-centred radicals into the surface of the 3D-printed object, which defuse to the surface and react with absorbed molecules to form covalent linkages long after the original PIII treatment. It has been shown that PIII can penetrate even small structures in a 3D-printed scaffold, providing an opportunity to create a 3D-printed scaffold that is attractive for patient cell colonisation [37, 39].

11.7 Bioprinting

The concept of 3D printing has also been applied to biology by using a CAD design to guide a pipette to layer (print) cells in a specific order to design and mimic a micro-environment of native tissue. The living cells in their culture medium are sometimes called bioink. The concept is essentially a sophisticated pipette, which places individual cells in the precise location that they would be found in a tissue, on the assumption they will mimic the native structure. Weekes *et al* [40] recently reviewed the use of bioprinting for small diameter vascular grafts, and highlight that one of the challenges of bioprinting is to maintain the structural integrity of the created constructs, which requires post-processing or hybrid material approaches. One of the few examples of bioprinting that has been translated into clinical practice is the 'BioPen', which is a handheld device that delivers two types of 'ink': the first ink uses cells in a biopolymer, and the second ink is a hydrogel that is cross-linked *in situ* using ultra-violet light. This device is designed to be used in surgery to repair bone and cartilage, where the surgeon can conduct free-form 3D sculpting to repair the anatomy [41].

Although there has been much hype about bioprinting, with the anticipation of printing functional whole organs, in reality the most useful application has been in the construction of tissue samples or of organoids, which can be used in scientific experiments. Organoids are 3D, self-assembled, vascularised, and functional tissue culture units that are derived from stem cells manufactured *in vitro* that mimic the function or dysfunction of an organ. Organoids differ from implantable microtissues by their organ-specific nature and self-assembling properties, and grow into viable microscopic versions of organs, such as lungs or pancreas. They are used for disease modelling research in the laboratory. Emerging clinical use for organoids includes radiation sensitivity, drug screening, the study of host-pathogen interaction, gene editing, biobanking, and ultimately patient-derived engineered constructs for regenerative medicine. One such example is the bioprinting of a kidney organoid, which has the full complement of cells that would be found in a kidney, which means that drug development and treatment response experiments can be performed on more complex systems than single cell lines [42]. The significance in radiation response studies will begin to emerge as these organoids replace single cell line *in vitro* studies.

Although animal models are the gold standard *in vivo* models for research and drug discovery, the majority of drugs that clear those preclinical animal trials fail in the clinical stage due to incompatibility issues between animals and humans [43]. Takebe *et al* [44] highlight that the failure of drug candidates can be explained by the reliance on mouse models and transformed cell lines. Organoids, boosted by advances in CRISP-R gene editing, have the potential to better mimic human physiology compared with animal models, and offer the possibility to improve toxicity and efficacy testing. Conversely, a 3D-printed organoid is isolated from all the factors to be found in a complete animal model. As the significance of the immune response and the microbiotic health of individuals on the response to drug and radiation treatments is emerging, the promise of bioprinting may turn out to be less than hoped for.

There is a long road ahead to step up from research laboratories to patient cell-derived organoid units for comprehensive drug screening, and ultimately implantable organs. Although there are several groups working with bioprinting, there are only a handful of organoid centres in the world focused on clinical translation, with emerging applications on intestinal, liver, cardiac, and lung organoids [44].

3D printing has near universal applicability in healthcare, and advances in tissue regeneration and bioprinting will increasingly integrate within existing models of care, such as anatomical models, surgical guides and prostheses. Ultimately 3D printing is one of many versatile tools, to be applied and combined with other approaches within the specific context of care to get the best possible outcome for each individual patient.

11.8 Key points

- 3D printing in medicine is a multidisciplinary endeavour that is reliant on collaboration between radiologists, treating clinicians, and design and manufacturing biomedical engineers and physicists.
- Practitioners should be familiar with their local regulatory framework for medical devices, regardless of which steps of the manufacturing workflow are performed at the point of care or outsourced. This includes an understanding of the quality and safety parameters associated with the software, hardware, materials, and credentials of the individuals involved in the production of the device.
- Anatomical models for surgical planning are currently the most reported use of 3D printing in medicine outside radiation therapy.
- Materials advances are offering new regenerative medicine approaches through bioresorbable polymers and tissue engineering. The emerging use of organoids may support evaluations of radiation sensitivity and drug screening.

References

[1] Diment L E, Thompson M S and Bergmann J H 2017 Clinical efficacy and effectiveness of 3D printing: a systematic review *BMJ Open* **7** e016891

[2] Vitali J, Cheng M and Wagels M 2019 Utility and cost-effectiveness of 3D-printed materials for clinical use *J. 3D Print. Med.* **3** 209–18

[3] Lee N 2019 Intellectual property rights and open innovation in 3D printing: A different form of exclusivity, in *The Innovation Society and Intellectual Property* (Cheltenham: Edward Elgar Publishing)

[4] Green N, Glatt V, Tetsworth K, Wilson L J and Grant C A 2016 A practical guide to image processing in the creation of 3D models for orthopedics *Tech. Orthopaedics* **31** 153–63

[5] Murray B, Desselle M, Forrestal D, Sparks L, Drummond S and Wagels M 2022 Three-dimensional printing in surgery: negotiating the regulatory landscape *ANZ J. Surg.* **92** 325–7

[6] Chen M Y, Skewes J, Woodruff M A and Rukin N J 2021 Using bespoke 3D-printed models to improve patient understanding of an encrusted ureteric stent *J. Clin. Urol* **14** 137–9

[7] Mathew N, Gandhi S, Singh I, Solanki M and Bedi N S 2019 3D models revolutionizing surgical outcomes in oral and maxillofacial surgery: experience at our center *J. Maxillofacial Oral Surg.* **19** 208–16

[8] Tetsworth K, Block S and Glatt V 2017 Putting 3D modelling and 3D printing into practice: virtual surgery and preoperative planning to reconstruct complex post-traumatic skeletal deformities and defects *SICOT-J* **3** 16

[9] Wong K C, Kumta S M, Geel N V and Demol J 2015 One-step reconstruction with a 3D-printed, biomechanically evaluated custom implant after complex pelvic tumor resection *Comput. Aided Surg.* **20** 14–23

[10] Randhawa K, West J, Skellern K and Josserand E 2021 Evolving a value chain to an open innovation ecosystem: cognitive engagement of stakeholders in customizing medical implants *Calif. Manage. Rev.* **63** 101–34

[11] Carter J C, Broadbent J, Murphy E C, Guy B, Baguley K E and Young J 2020 A three-dimensional (3d) printed paediatric trachea for airway management training *Anaesthesia Intensive Care* **48** 243–5

[12] Pang G, Futter C, Pincus J, Dhanani J and Laupland K 2020 Development and testing of a low cost simulation manikin for extracorporeal cardiopulmonary resuscitation (ECPR) using 3-dimensional printing *Resuscitation* **149** 24–9

[13] Ballard D H, Mills P, Duszak R Jr, Weisman J A, Rybicki F J and Woodard P K 2020 Medical 3D printing cost-savings in orthopedic and maxillofacial surgery: cost analysis of operating room time saved with 3D printed anatomic models and surgical guides *Acad. Radiol.* **27** 1103–13

[14] Bauer A S, Storelli D A, Sibbel S E, McCarroll H R and Lattanza L L 2017 Preoperative computer simulation and patient-specific guides are safe and effective to correct forearm deformity in children *J. Pediatric Orthopaedics* **37** 504–10

[15] Sharma N, Cao S, Msallem B, Kunz C, Brantner P, Honigmann P and Thieringer F M 2020 Effects of steam sterilization on 3D printed biocompatible resin materials for surgical guides–an accuracy assessment study *J. Clin. Med.* **9** 1506

[16] Design With Us 2023 *Our Origin Story* [Online] Available: https://www.designwithus.org/our-origin-story (visited on 05/12/2023)

[17] Enabling The Future 2023 *Enabling the Future—A Global Network of Passionate Volunteers Using 3D Printing to Give the World a "helping hand."* [Online] Available: https://enablingthefuture.org/ (visited on 05/12/2023)

[18] Wendo K, Barbier O, Bollen X, Schubert T, Lejeune T, Raucent B and Olszewski R 2022 Open-source 3D printing in the prosthetic field–the case of upper limb prostheses: a review *Machines* **10** 413

[19] Manero A, Smith P, Sparkman J, Dombrowski M, Courbin D, Kester A, Womack I and Chi A 2019 Implementation of 3D printing technology in the field of prosthetics: Past, present, and future *Int. J. Environ. Res. Public Health* **16** 1641

[20] Kim S, Yalla S, Shetty S and Rosenblatt N J 2022 3D printed transtibial prosthetic sockets: a systematic review *PLoS One* **17** e0275161

[21] Wojciechowski E, Chang A Y, Balassone D, Ford J, Cheng T L, Little D, Menezes M P, Hogan S and Burns J 2019 Feasibility of designing, manufacturing and delivering 3D printed ankle-foot orthoses: a systematic review *J. Foot Ankle Res.* **12** 11

[22] Waldburger L, Schaller R, Furthmüller C, Schrepfer L, Schaefer D J and Kaempfen A 2021 3D-printed hand splints versus thermoplastic splints: a randomized controlled pilot feasibility trial *Int. J. Bioprint* **8** 474

[23] Puls N, Carluccio D, Batstone M D and Novak J I 2021 The rise of additive manufacturing for ocular and orbital prostheses: a systematic literature review *Ann. 3D Print. Med.* **4** 100036
[24] McKittrick A, Brown J, Desselle M, Forrestal D, Fakhr R and Skewes J 2022 *3D Printed Transparent Facial Orthosis—A Patient's Experience* [Online] Available: https://anzbaasm.com/8260 (visited on 05/12/2023)
[25] iOrthotics 2021 *Our Difference* [Online] Available: https://www.iorthotics.com.au/our-difference/ (visited on 05/12/2023)
[26] Cruz R L, Ross M T, Skewes J, Allenby M C, Powell S K and Woodruff M A 2020 An advanced prosthetic manufacturing framework for economic personalised ear prostheses *Sci. Rep.* **10** 11453
[27] Tetsworth K, Woloszyk A and Glatt V 2019 3D printed titanium cages combined with the masquelet technique for the reconstruction of segmental femoral defects *OTA Int.* **2** e016
[28] Jaggessar A, Shahali H, Mathew A and Yarlagadda P K 2017 Bio-mimicking nano and micro-structured surface fabrication for antibacterial properties in medical implants *J. Nanobiotechnol.* **15** 64
[29] Kobbe P, Laubach M, Hutmacher D W, Alabdulrahman H, Sellei R M and Hildebrand F 2020 Convergence of scaffold-guided bone regeneration and RIA bone grafting for the treatment of a critical-sized bone defect of the femoral shaft *Eur. J. Med. Res.* **25** 70
[30] Cheng M E, Janzekovic J, Theile H J, Rutherford-Heard C, Wille M-L, Cole C, Lloyd T B, Theile R J, Wagels M and Hutmacher D W 2022 Pectus excavatum camouflage: a new technique using a tissue engineered scaffold *Eur. J. Plast. Surg.* **45** 177–82
[31] Queensland Government 2022 *Advance Queensland Research Leads to World-First Breast Implant Surgery Using 3D Printed Scaffold* [Online] Available: https://advance.qld.gov.au/whats-happening/stories-about-innovation/3d-printed-scaffold (visited on 05/12/2023)
[32] Shafiee A, Cavalcanti A S, Saidy N T, Schneidereit D, Friedrich O, Ravichandran A, De-Juan-Pardo E M and Hutmacher D W 2021 Convergence of 3D printed biomimetic wound dressings and adult stem cell therapy *Biomaterials* **268** 120558
[33] Wang P-J, Ferralis N, Conway C, Grossman J C and Edelman E R 2018 Strain-induced accelerated asymmetric spatial degradation of polymeric vascular scaffolds *Proc. Natl. Acad. Sci.* **115** 2640–5
[34] Yazici G *et al* 2016 The dosimetric impact of implants on the spinal cord dose during stereotactic body radiotherapy *Radiat. Oncol.* **11** 1–9
[35] Kairn T, Crowe S, Kenny J, Mitchell J, Burke M, Schlect D and Trapp J 2013 Dosimetric effects of a high-density spinal implant *J. Phys.: Conf. Ser.* **444** 012108
[36] Katsifis G A, McKenzie D R, Hill R, Connor M O, Milross C and Suchowerska N 2022 Radiation dose perturbation at the tissue interface with PEEK and titanium bone implants: Monte Carlo simulation, treatment planning and film dosimetry *Radiat. Phys. Chem.* **199** 110398
[37] Katsifis G A, McKenzie D R and Suchowerska N 2019 Applying the Hashin-Shtrikman bounds to predict stiffness of multicomponent 3D printed structures: towards regenerative orthopaedic medicine *J. Compos. Mater.* **54** 2173–83
[38] Kruse H V, McKenzie D R, Clark J R and Suchowerska N 2021 Plasma ion implantation of 3D-printed PEEK creates optimal host conditions for bone ongrowth and mineralisation *Plasma Process. Polym.* **18** 2000219

[39] Katsifis G, Kruse H, Lewin W, Al Maruf A, Clark J R, McKenzie D R and Suchowerska N 2023 Micro-CT analysis of implanted poly-ether-ether-ketone scaffolds: plasma immersion ion implantation increases osteoconduction *Adv. Eng. Mater.* **25** 2201297

[40] Weekes A, Bartnikowski N, Pinto N, Jenkins J, Meinert C and Klein T J 2022 Biofabrication of small diameter tissue-engineered vascular grafts *Acta Biomater.* **138** 92–111

[41] O'Connell C D *et al* 2020 Free-form co-axial bioprinting of a gelatin methacryloyl bio-ink by direct *in situ* photo-crosslinking during extrusion *Bioprinting* **19** e00087

[42] Phipson B *et al* 2019 *Nat. Methods* **16** 79–87

[43] Choudhury D, Ashok A and Naing M W 2020 Commercialization of organoids *Trends Mol. Med.* **26** 245–9

[44] Takebe T, Wells J M, Helmrath M A and Zorn A M 2018 Organoid center strategies for accelerating clinical translation *Cell Stem Cell* **22** 806–9

Chapter 12

Conclusions

Tanya Kairn, Scott B Crowe and Tomas Kron

During the last decade, 3D printing has become an integral part of many industrial applications. The ability to rapidly realise physical objects of many shapes, materials, and sizes supports a 'just in time' production cycle and supplies an increasing need to find customised solutions. Additionally, 3D-printing techniques enable cost-effective production of unique objects or small numbers of identical items. This small-scale production capability is highly advantageous in medical applications, including radiation therapy (as described in this book).

Being founded on the basis of 'first do no harm' (primum non nocere), this book has a considerable focus on ensuring the safe use of 3D printing in radiation therapy through good planning, verification, record keeping, and other quality management processes. Despite this weight of responsibility, it is important to understand that the implementation of a clinical 3D-printing service in a radiation therapy centre can still be simple, efficient, inexpensive, and comparatively fast.

A straightforward road to clinical implementation of a 3D-printing service involves starting with one printer, one material, and one clinical problem to solve, and an overall process, such as:

- Work with a multidisciplinary team to identify the clinical need that is to be filled through 3D printing (see examples in chapters 9 and 10);
- Select and acquire the 3D printer (see chapter 2) and material (see chapter 3) that will be used;
- Devise a method for producing the 3D-printed devices (see examples in chapters 4 and 5);
- Work with a multidisciplinary team to ensure that the workflow for producing the 3D prints is understood and agreed, and that the 3D-printing work will be supported by sufficient staffing and physical space to ensure safety (see chapters 6 and 7);
- Investigate and document risks associated with the 3D-printing service (see chapters 7 and 8), including risks to patients (see chapter 9);

doi:10.1088/978-0-7503-3907-0ch12

- Complete a commissioning process for the printer and material (covering print parameter selection, sample print testing, and development of print sanitisation processes, as well as quality control processes for the printer and the prints, see chapter 8);
- Complete all required staff training and process documentation prior to clinical release of the printer and material (see chapter 8); and
- Complete periodic reviews of quality control results and logs, and seek staff and patient feedback regarding the 3D-printing service and the clinical use of the 3D-printed devices, to ensure ongoing quality improvement (see chapters 7 and 8).

It is entirely possible to produce excellent results that improve the patient and staff experience of the radiation therapy treatment process, by just starting small and keeping good records. Completing a comprehensive and well-documented commissioning process for the first 3D-printer and material will make the commissioning of all subsequent printers and materials that much easier.

Obviously, the advice provided in this book should be considered as additional to local regulatory requirements—the requirements of all relevant laws and standards that apply to the jurisdiction in which the radiation therapy centre is located must be followed. For example, as noted in chapter 7, some jurisdictions require registration of the manufacture of medical devices, including medical devices that are made in-house, and therefore such registration must be sought for 3D-printed radiation therapy devices in those jurisdictions. However, even if not required by regulations, 3D-printed radiation therapy devices must be assessed for safety in accordance with their intended application (e.g. sanitisation for devices to be used in contact with patients, strength for devices to be used as supports, etc), and the radiation attenuation and scattering effects of any devices placed in or close to the radiation beam must be assessed to ensure treatment accuracy.

An appropriately managed, resourced, and supported 3D-printing service can be a net benefit to the radiation therapy centre in terms of patient experience, staff engagement, treatment accuracy, and improved treatment outcomes, in addition to monetary costs. As described in chapter 11, the ongoing development of processes for 3D printing in radiation therapy exists alongside research into 3D printing for a range of cutting-edge medical applications, and the opportunities to collaborate and learn from each other's work are manifold.

The content of this book is based on the existing scientific literature at the time of writing, as well as on the breadth of experience of the various chapter authors, much of which has also relied on early and frequent re-examination of information in the scientific literature. The field of 3D printing in radiation therapy, like so much of medicine, requires the liberal dissemination of ideas, data, and experience. For this reason, it is our heartfelt intention that the readers of this book will use the information herein to build, extend, and enhance their own 3D-printing services, and will add their experiences to the scientific literature, for the benefit of all of us and our patients.

List of acronyms and initialisms

1D	one-dimensional
1SD	one standard deviation from the mean of a sample
2D	two-dimensional
2PL	two-photon polymerisation (also known as TPP, see chapter 2)
3D	three-dimensional
4D	four-dimensional
ABS	acrylonitrile butadiene styrene (see chapter 3)
AM	additive manufacture (a common name for 3D printing, see chapter 2)
BJT	binder jetting (see chapter 2)
CAD	computer aided design (see chapter 4)
CAM	computer aided manufacturing (see chapter 5)
CAM-LEM	computer-aided manufacturing of laminated engineering materials (see chapter 2)
CBAM	composite based additive manufacturing (see chapter 2)
CDLP	continuous direct light processing (also known as continuous liquid interface production, see chapter 2)
ClinRO	clinician reported outcome (see chapter 11)
CNC	computer numerical control (see chapter 5)
CT	computed tomography
DCP	dual-colour photo-polymerisation (see chapter 2)
DED	direct energy deposition (see chapter 2)
DfMA	design for manufacture and assembly (see chapter 7)
DICOM	digital imaging and communications in medicine (file format, see chapter 4)
DIW	direct ink writing (see chapter 2)
DLP	digital light processing (see chapter 2)
DMLS	direct metal layer sintering (see chapter 2)
DOD	drop-on-demand (see chapter 2)
EBAM	electron beam additive manufacturing (see chapter 2)
EBM	electron beam melting (see chapter 2)
EBRT	external beam radiation therapy (see chapters 1, 9 and 10)
ECMO	extracorporeal membrane oxygenation (see chapter 11)

doi:10.1088/978-0-7503-3907-0ch13

ECPR	extracorporeal cardiopulmonary resuscitation (see chapter 11)
EHD	electrohydrodynamic (see chapter 2)
FDG	fluorodeoxyglucose (a PET tracer, see chapter 10)
FDM	fused deposition modelling (commercial name for MEX or FFF, see chapter 2)
FFF	fused filament fabrication (alternative name for MEX or FDM, see chapter 2)
FMEA	failure modes and effects analysis (see chapter 7)
g-code	geometry code (3D printer instructions, see chapter 5)
GMDN	Global Medical Device Nomenclature (see chapter 7)
GPU	graphical processing unit (see chapter 4)
HDR	high dose rate (type of brachytherapy, see chapters 1 and 9)
HDT	heat distortion temperature (see chapter 5)
HIPS	high impact polystyrene (see chapter 3)
HU	Hounsfield unit
IEC	International Electrotechnical Commission (see chapter 7)
IGRT	Image-guided radiation therapy (see chapter 10)
IMRT	intensity modulated radiation therapy (see chapters 8, 9 and 10)
IPA	isopropyl alcohol (solvent, see chapter 5)
ISO	International Standards Organisation (see chapter 7)
kV	kilovoltage/kilovolts
LDR	low dose rate (type of brachytherapy, see chapters 1 and 9)
LENS	laser engineering net shape (see chapter 2)
LIFT	laser induced forward transfer (see chapter 3)
Linac	linear accelerator
LMPA	low melting point alloy (see chapter 9)
LOM	laminated object manufacturing (see chapter 2)
MBDCA	model-based dose calculation algorithm (see chapter 9)
MDT	multi-disciplinary team (see chapter 7)
MeV	mega electron voltage/mega electron volts
MEW	melt-electrowriting (see chapter 2)
MEX	material extrusion (also known as FDM or FFF, see chapter 2)
MJ	material jetting (see chapter 2)
MJF	multi jet fusion (see chapter 2)
MJP	multi-jet printing (also known as MJT, see chapter 2)
MJT	material jetting (also known as MJP, see chapter 2)
MRI	magnetic resonance imaging (see chapters 4 and 10)
MSDS	materials safety data sheet (see chapter 8)
MV	megavoltage/megavolts
NPJ	nanoparticle jetting (see chapter 2)
PA, PA11, PA12, PA2200, PA3200	common polyamides (see chapter 3)
PACS	picture archiving and communication system (see chapter 4)
PE	polyethylene (see chapter 3)
PEI	polyetherimide (a print bed material, see chapter 2)
PBF	powder bed fusion (see chapter 2)
PC	polycarbonate (see chapter 3)
PDD	percentage depth dose (see chapter 9)
PEEK	polyether ether ketone (see chapter 3)
PEK	polyetherketone (see chapter 11)

PEKK	polyetherketoneketone (see chapter 11)
PET	positron emission tomography
PETG	polyethylene terephthalate glycol (see chapter 3)
PIII	plasma immersion ion implantation (see chapter 11)
PLA	polylactic acid (see chapter 3)
PLA-G-CNF	PLA-graphene carbon nanofibre (see chapter 3)
PMMA	polymethyl methacylate (see chapter 3)
PμSL	projection micro stereolithography (see chapter 2)
PPE	personal protective equipment (see chapter 6)
PPSF	polyphenylsulfone (see chapter 3)
PROM	patient reported outcome measure (see chapter 11)
PSL	plastic sheet lamination (see chapter 2)
PVA	polyvinyl alcohol (see chapter 3)
PVC	polyvinyl chloride (see chapter 3)
QMS	quality management system (see chapter 7)
RED	relative electron density (see chapter 8)
SDL	selective deposition lamination (see chapter 2)
SGRT	surface-guided radiation therapy (see chapters 8 and 10)
SHL	sheet lamination (see chapter 2)
SLCOM	selective lamination composite object manufacturing (see chapter 2)
SLA	stereolithography (see chapter 2)
SLM	selective layer melting (see chapter 2)
SLS	selective layer sintering (see chapter 2)
SPECT	single photon emission computed tomography
SSD	source to surface distance (see chapter 10)
STL	stereolithography (file format for 3D models, also known as 'standard tessellation language' and 'standard triangle language', see chapter 4)
TMR	tissue-maximum ratio (see chapter 9)
TPE	thermoplastic elastomer (see chapter 3)
TPP	two-photon polymerisation (also known as 2PL, see chapter 2)
TPS	treatment planning system
TPU	thermoplastic polyurethane (see chapter 3)
TQM	total quality management (see chapter 7)
TSET, TSEI, or TSEBT	total skin electron radiation therapy (see chapter 9)
VPP	vat photopolymerisation (see chapter 2)
UAM	ultrasonic additive manufacturing (see chapter 2)
UFP	ultrafine particle (see chapter 6)
UTS	ultimate tensile strength (see chapter 5)
VMAT	volumetric modulated arc therapy (see chapters 8, 9 and 10)
VOC	volatile organic compound (see chapter 6)

www.ingramcontent.com/pod-product-compliance
Ingram Content Group UK Ltd.
Pitfield, Milton Keynes, MK11 3LW, UK
UKHW051854100926
8556IPUK00005B/622

9 780750 33905